A Knowledge Management Approach to Drinking Water Utility Business

Awwa Research Foundation

About the Awwa Research Foundation

The Awwa Research Foundation (AwwaRF) is a member-supported, international, nonprofit organization that sponsors research to enable water utilities, public health agencies, and other professionals to provide safe and affordable drinking water to consumers.

The Foundation's mission is to advance the science of water to improve the quality of life. To achieve this mission, the Foundation sponsors studies on all aspects of drinking water, including supply and resources, treatment, monitoring and analysis, distribution, management, and health effects. Funding for research is provided primarily by subscription payments from approximately 1,000 utilities, consulting firms, and manufacturers in North America and abroad. Additional funding comes from collaborative partnerships with other national and international organizations, allowing for resources to be leveraged, expertise to be shared, and broad-based knowledge to be developed and disseminated. Government funding serves as a third source of research dollars.

From its headquarters in Denver, Colorado, the Foundation's staff directs and supports the efforts of more than 800 volunteers who serve on the board of trustees and various committees. These volunteers represent many facets of the water industry, and contribute their expertise to select and monitor research studies that benefit the entire drinking water community.

The results of research are disseminated through a number of channels, including reports, the Web site, conferences, and periodicals.

For subscribers, the Foundation serves as a cooperative program in which water suppliers unite to pool their resources. By applying Foundation research findings, these water suppliers can save substantial costs and stay on the leading edge of drinking water science and technology. Since its inception, AwwaRF has supplied the water community with more than $300 million in applied research.

More information about the Foundation and how to become a subscriber is available on the Web at **www.awwarf.org.**

A Knowledge Management Approach to Drinking Water Utility Business

Prepared by:
David M. Smigiel, **John G. Sulewski**, and **Michael A. Moss**
Westin Engineering
407 East Fort Street, Suite 200, Detroit, MI 48226-2900

Sponsored by:
Awwa Research Foundation
6666 West Quincy Avenue, Denver, CO 80235-3098

Published by:

Awwa Research Foundation

American Water Works Association

IWA Publishing

DISCLAIMER

This study was funded by the Awwa Research Foundation (AwwaRF). AwwaRF assumes no responsibility for the content of the research study reported in this publication or for the opinions or statements of fact expressed in the report. The mention of trade names for commercial products does not represent or imply the approval or endorsement of AwwaRF. This report is presented solely for informational purposes.

CONTENTS

TABLES

FIGURES

FOREWORD

The Awwa Research Foundation is a nonprofit corporation that is dedicated to the implementation of a research effort to help utilities respond to regulatory requirements and traditional high-priority concerns of the industry. The research agenda is developed through a process of consultation with subscribers and drinking water professionals. Under the umbrella of a Strategic Research Plan, the Research Advisory Council prioritizes the suggested projects based upon current and future needs, applicability, and past work; the recommendations are forwarded to the Board of Trustees for final selection. The foundation also sponsors research projects through the unsolicited proposal process; the Collaborative Research, Research Applications, and Tailored Collaboration programs; and various joint research efforts with organizations such as the U.S. Environmental Protection Agency, the U.S. Bureau of Reclamation, and the Association of California Water Agencies.

This publication is a result of one of these sponsored studies, and it is hoped that its findings will be applied in communities throughout the world. The following report serves not only as a means of communicating the results of the water industry's centralized research program but also as a tool to enlist the further support of the nonmember utilities and individuals.

Projects are managed closely form their inception to the final report by the foundation's staff and large cadre of volunteers who willingly contribute their time and expertise. The foundation serves a planning and management function and awards contracts to other institutions such as water utilities, universities, and engineering firms. The funding for this research effort comes primarily from the Subscription Program, through which water utilities subscribe to the research program and make an annual payment proportionate to the volume of water they deliver and consultants and manufacturers subscribe based on their annual billings. The program offers a cost-effective and fair method for funding research in the public interest.

A broad spectrum of water supply issues is addressed by the foundation's research agenda: resources, treatment and operations, distribution and storage, water quality and analysis, toxicology, economics, and management. The ultimate purpose of the coordinated effort is to assist water suppliers to provide the highest possible quality of water economically and reliably. The true benefits are realized when the results are implemented at the utility level. The foundation's trustees are pleased to offer this publication as a contribution toward that end.

Walter J. Bishop
Chair, Board of Trustees
Awwa Research Foundation

Robert C. Renner, P.E.
Executive Director
Awwa Research Foundation

ACKNOWLEDGMENTS

The authors of this report are indebted to the following utilities and individuals for their cooperation, funding, and participation in this project:

Anchorage Water and Wastewater Utility, Anchorage, Alaska, Mark C. Premo
Austin Water Utility, Austin, Texas, Brownlee Bowmer
Chicago Department of Water Management, Chicago, Ill., John Spatz
City of Cleveland Division of Water, Cleveland, Ohio, Ron Petrich
Columbus Water Works, Columbus, Ga., Cliff Arnett
County of Henrico Department of Public Utilities, Henrico County, Va., Arthur D. Petrini
Dallas Water Utilities Department, Dallas, Texas, Robert M. Johnson
Detroit Water and Sewerage Department, Detroit, Mich., Kevin Quiggle
Greater Cincinnati Water Works, Cincinnati, Ohio, Steven Hellman
Lafayette Utilities System, Lafayette, La., Don Broussard
Las Vegas Valley Water District, Las Vegas, Nev., James W. Ellisor
Minneapolis Water Works, Minneapolis, Minn., Adam J. Kramer
Oakland County Drain Commission, Waterford, Mich., Kevin R. Larsen
Portland Water Bureau, Portland, Ore., Mark Knudson
Portland Water District, Portland, Maine, Gordon Johnson
San Diego County Water Authority, San Diego, Calif., Marilyn Young and Shelli Ellison

The advice and help of the Awwa Research Foundation senior project manager (Linda Reekie) and knowledge manager (Jeff Oxenford) – as well as the Project Advisory Committee (Jim Lee, APQC, Cleveland, Ohio; Audrey A. Cudrak, EPCOR Water Services, Edmonton, Alta.; Dave Muto, Seattle Public Utilities, Seattle, Wash.; and Linda Godfrey, Thames Water Utilities, London, United Kingdom) – is especially noted with much appreciation. Their commitment to this project and to the industry has been exemplary.

The authors express their appreciation for the participation of Richard Stevens, National Aeronautics and Space Administration (NASA), Kennedy Space Center, Fla., whose insights into how a Knowledge Management program is working outside the drinking water industry proved stimulating and invaluable.

Special recognition is extended to the American Water Works Association (AWWA) for their coordinating assistance with the on-site workshop held prior to its 2004 annual conference.

The Association of Metropolitan Water Agencies (AMWA) graciously allowed the inclusion of a Knowledge Management presentation and workshop during its 2004 annual meeting. This opportunity validated the research results of this project, and the authors are most grateful.

The authors wish to acknowledge the contributions of Doug Harp, Thomas J. DeLaura, Ed Hogan, Prem Jha, and Ken Smaellie, all from Westin Engineering.

PREFACE

> The dogmas of the quiet past are inadequate to the stormy present. The occasion is piled high with difficulty, and we must rise to the occasion. As our case is new, so must we think anew and act anew. (Abraham Lincoln)

That the drinking water utility sector is moving toward a business-centric view of their operations beyond the treatment and pumping of safe drinking water is evident. This shift was demonstrated and related during the Awwa Research Foundation's Knowledge Management (KM) project on numerous occasions, as exemplified by:

- Utilities grappling with the loss of knowledgeable senior personnel and struggling to train and recruit a "new" workforce. One executive related his personal challenge of potentially losing his top and middle operations management team within 9 months to an early retirement offer.
- Directors committing their organizations to executive management training such as the Water and Wastewater Leadership Center (Kenan-Flagler Business School, University of North Carolina) and creating and/or realigning management positions to optimize operations by integrating and streamlining business processes and information systems.
- Consistent emphasis across the project team regarding the identification of key performance metrics to guide operations. Metrics beyond treating water including customer satisfaction, workforce recruitment, training, and public relations.
- A firm and passionate belief that their organizations need to move toward becoming "learning organizations" as a way to more effectively address the dynamics of the regulatory, workforce, and business environments.

The combination of critical pressures such as a smaller and less-experienced workforce, ever-questioning constituency, reduced budgets, increased regulations, heightened sensitivity toward security, and aging infrastructure are redefining the role of Utility Managers. No longer can Managers take for granted that they have the time (schedule overruns), money (rework), resources, infrastructure and expertise (institutional knowledge) to meet this new landscape of challenges. To be successful in meeting the challenges of this current business climate, Utility Managers must find innovative ways to leverage organizational capabilities and past experiences in designing new strategies to address this present set of hurdles.

Echoed by many, one project participant offered, "We have to do a better job at recognizing, capturing, and protecting our digital assets just like our infrastructure assets." The belief that one's individual utility was not doing a good enough job capturing, coordinating, and integrating its data and information was consistent across the project team. Other consistent themes included the perceived value of qualified key performance indicators and the impact from workforce attrition and related loss of "know-how."

The Utility Manager must recognize that a key strategy to operating in today's business climate involves capturing, managing, integrating and applying institutional knowledge (experience, know-how, and expertise). KM is a business strategy and core competency that utilities must develop to operate with a smaller, more flexible workforce, respond to public scrutiny, "do more with less," meet stiffening regulations, protect assets, and maximize the useful life of infrastructure.

Although Capital Improvement Projects may be required to support KM, importantly KM should not be treated as a Capital Improvement Project with a defined beginning and end. It must be viewed as an organizational evolution and journey encompassing people, culture, processes, and technology. For KM strategies to be successful, they must maintain an alignment with business objectives; KM initiatives must be viewed as strategic with properly skilled resources allocated and ownership assigned; and measurement of results must be incorporated. Philosophically, KM strategies must be discussed throughout the organization and embedded into each initiative the utility undertakes. If viewed as "business as usual," over time the organization will realize an enhanced capacity to innovate and adapt to the ever-changing business climate.

For justification, the Utility Manager need look no further than his/her peers in the automotive, pharmaceutical, defense, high technology, and energy industries and government agencies who have embraced KM not only as a means to increase efficiency, but also for realizing and sustaining competitive advantage.

Unlike many of these other industries, the drinking water industry embraces and promotes the notion of a collaborative community to share experiences, lessons learned, best practices, and solutions for the betterment of all of its members. The existing infrastructure of associations, agencies, and a world-class research foundation, coupled with a vibrant and mature volunteer community, provide the mechanisms necessary for coordination, experimentation, and diffusion of KM-based strategies, standards, and solutions applicable to all utilities, regardless of their individual characteristics. This unique model of cooperation positions the industry to rapidly learn from KM practitioners and to adapt their lessons learned and best practices to the unique characteristics of the utility organization and environment in which it functions.

In summary, this project confirmed the applicability of KM strategies to the current industry business environment. Embraced and validated by other industries, KM has been shown to be successful in addressing challenges similar to those that the drinking water industry is facing. The culture of the industry provides a fertile ground for the development of strategies, standards and solutions from which all utilities can benefit. A critical result of this project is the development of follow-up candidate research projects that will continue the momentum and further develop KM as a mainstream management strategy for the industry. Most importantly, the active engagement of the project team and high degree of curiosity pertaining to the project – as demonstrated by invitations to speak at meetings of the Association of Metropolitan Water Agencies (AMWA), various state sections and local chapters, Association of Metropolitan Sewerage Agencies (AMSA)*, and the AWWA/WEF Joint Management Conference – underline the industry's interest in learning about and applying this compelling management strategy.

*AMSA officially changed its name to the National Association of Clean Water Agencies (NACWA) in May 2005.

EXECUTIVE SUMMARY

> Exploiting the informated environment means opening the information base of the organization to members at every level, assuring that each has the knowledge, skills and authority to engage with the information productively. (S. Zuboff, *Scientific American*)

BACKGROUND

Initiatives with objectives of identifying, managing, and maintaining knowledge within organizations have been commonplace outside of the drinking water utility sector (the industry) for more than a decade. Many of these organizations have become thought leaders on how to leverage Knowledge Management (KM) strategies as part of solutions to their business challenges. This key strategic management trend is providing leading organizations and agencies from pharmaceutical, automotive, high-technology, defense, energy, construction, and government arenas with capabilities for mitigating the loss of corporate knowledge. Additionally, while utilizing KM strategies, these organizations have identified sound approaches to managing the volumes of data and information that exist and in using this information efficiently and effectively for improving decision-making ability. The personnel in these organizations have become more effective as a result of these strategies due to enhanced horizontal communication pathways across the functional boundaries of the organization. In many of these situations, this improved peer-to-peer collaboration has been a catalyst for establishing and maintaining a knowledge-enabled environment. Additionally, significant gains relative to interactions with key external stakeholders such as customers, suppliers, and governing bodies have also been achieved.

KM practitioners and scholars continue to develop the body of knowledge documenting the justification for making the management of corporate information assets a strategic objective along with best practices for implementing KM initiatives. Studying the lessons learned from these early adopters and applying the business-oriented collaborative management principles inherent in KM can assist the industry in becoming more efficient, effective, proactive, and adaptive.

The Awwa Research Foundation (AwwaRF) recognized the applicability of the business-oriented collaborative management principles inherent in KM to the water industry in order to enable more effective use of organizational data and information. Toward this end, in late 2003 AwwaRF sponsored Project 2933, "A Knowledge Management Approach to the Water Industry." This report summarizes the findings of this research work that was started in late 2003 and generally concluded at the end of 2004.

OBJECTIVES AND GOALS

The overarching objective of this project is to begin the development of and establish a mechanism for maintaining and evolving an industry environment that places a high value on a learning and innovative culture. The primary goals of this project are to identify KM's applicability to the key challenges of the industry, create a working definition of KM for the industry, develop a working "best practice" bibliography highlighting case studies from outside the industry, and identify 5 to 10 additional candidate research projects that would further define the application of KM for the industry.

The development of a working definition for KM was a critical step in uniting the project participants around a common purpose. Reaching this consensus established the beginnings of a paradigm which can be further evolved by critical-thinking leaders and strategists within the industry. The current working definition combines themes found in definitions of KM practitioners from outside the industry with industry-specific criteria. As a result of the project team's increased insight and awareness of the KM domain, the definition was modified several times through the course of the project and is currently described as follows:

KM is a business strategy by which a water utility consciously identifies, captures, indexes, manages, and stores experiences, data, and information and provides methods for easily accessing and acting upon these collective assets in a collaborative environment optimizing the use of people, processes, and technology in support of:

- *Effective decision making*
- *Assuring compliance*
- *Improving performance*
- *Innovation*
- *Business continuity, all on a timely and sustainable basis.*

Although most project participants represented public utilities, private utilities were also represented. To comprehend the business model for private organizations, a modified definition is suggested which would include the addition of a new final sentence:

KM is a business strategy by which a water utility consciously identifies, captures, indexes, manages, and stores experiences, data, and information and provides methods for easily accessing and acting upon these collective assets in a collaborative environment optimizing the use of people, processes, and technology in support of:

- *Effective decision making*
- *Assuring compliance*
- *Improving performance*
- *Innovation*
- *Business continuity, all on a timely and sustainable basis.*

KM embodies processes by which the organization creates wealth (i.e., profit) by leveraging its intellectual and knowledge-based assets.

An investment in knowledge always pays the best interest. (Benjamin Franklin)

INDUSTRY BENEFITS

This project provides the following benefits to the drinking water industry:

- Initiates the development of a common language and framework for the application of KM
- Establishes a foundation for developing an understanding of KM's applicability with a community of industry leaders
- Surfaces examples of "best-in-class" KM applications both from outside and within the industry
- Identifies industry business drivers that are closely aligned with the drivers of other industries that have leveraged KM to improve their business models
- Begins to develop an understanding of where KM strategies may be of value in overcoming the current set of industry challenges
- Provides a framework for additional research considerations by identifying several research candidate projects to further refine KM's applicability and value to the industry.

APPROACH

The project approach centered on recruiting key stakeholders representative of the industry's demographics as participants in a community focused on learning current perspectives pertaining to KM. This community shepherds the thought leadership necessary to realizing meaningful applications of KM for the industry. The primary method for the community members to learn and collaborate was a project team Web site consisting of KM-enabling tools such as document management and discussion databases. Utilizing this Web site, virtual working sessions were held to discuss and learn current perspectives relating to KM implementations and applications for solving business problems. These virtual sessions were instrumental as they facilitated collaboration and active participation of the team without requiring the time and expense of travel. Another key tool was Web-based surveys (refer to Appendix A) that the participants completed at several milestones during the project. These surveys collected background with respect to the participants and their utility relative to the key ingredients of KM: people, culture, process, and technology.

In concert with the June 2004 annual AWWA conference in Orlando, Florida, an on-site workshop was conducted for all of the participants. During this workshop several team members presented examples of initiatives addressing critical business issues currently underway at their utilities. Also, to provide a perspective from outside the industry, two organizations prominent in the study and application of KM – the National Aeronautics and Space Administration (NASA) and APQC – were asked to present their experiences with KM. These presentations proved quite beneficial in that they highlighted practical applications of KM and provided insight into how other practitioners are addressing common business challenges through the application of KM strategies. A key exercise conducted during this on-site meeting was a facilitated brainstorming session during which the team members identified ideas for utilizing KM strategies to assist with solving industry problems. These ideas served as the foundation for the candidate research projects described below.

Successful knowledge transfer involves neither computers nor documents, but rather interactions between people. (T.H. Davenport, *CIO Magazine*)

RECOMMENDATIONS FOR FUTURE RESEARCH

The research team, in collaboration with the project participants, identified the following candidates for future research projects. These candidates have common objectives of further defining the application of KM for the industry and identifying best practices and practical approaches for the inclusion of KM within the mainstream of utility management strategies. In this section the candidate research project's recommended sequence of execution and a brief statement pertaining to each project objective are presented. Additional detail regarding the project profiles may be found in Section VII, Candidate Research Projects.

Rationale for Candidate Project Sequencing

Although the understanding of KM strategies is relatively new to the industry as a whole, the research did identify that several of the participating utilities are attempting to strategically utilize concepts found in the KM domain when approaching solutions to solving their data, information, and knowledge management challenges. Additionally, a few of the participating utilities are actively addressing knowledge transfer and retention issues in trying to deal with the "brain drain" that is occurring as a result of the workforce dynamics. Examples of this are the interview sessions for key departing employees at Thames Water Utilities, succession planning focus done at Anchorage Water and Wastewater Utility, and activities (e.g., continuous improvement strategies, digitizing coded knowledge) at other utilities (e.g., Columbus Water Works, Greater Cincinnati Water Works, and Minneapolis Water Works) representing dimensions of KM. Outside of the project team, through validation activities such as AMWA workshops, the research team confirmed that similar activities are occurring at a significant number of other utilities.

Developed from a practitioner's viewpoint, the sequencing recommendation for the candidate research projects is one that provides mechanisms for the industry to realize benefits and applicable lessons learned in as quick a time as possible while in parallel establishing a formal body of knowledge over time that will result in practical knowledge management strategies, standards, and solutions.

Figure ES.1 depicts the candidate research projects in terms of an execution sequence. It also portrays their contribution to the formulation of industry strategies, standards, and solutions to the development of Knowledge Management as a mainstream management strategy for the industry. In this context, *strategy* is defined as an approach that will affect the overall direction of the KM program and will establish the utility's future environment with respect to launching and sustaining KM initiatives. *Standard* is defined as a criterion or model established by general consent. Lastly, *solution* is a tool that facilitates resolution of a specific business issue and typically incorporates an application of technology. Here *technology* is defined to include at a minimum hardware, software, and processes necessary to facilitate a resolution.

The logic for the recommended sequence portrayed in Figure ES.1 utilized the following criteria:

- Identification of a KM dimension to assist with solving a current critical industry problem
- Importance to launching and sustaining a KM strategy
- Contribution of building block standards for the industry's KM program
- Specification of how a utility with limited technology investment can benefit from KM.

The recommended sequence as shown in Figure ES.1 follows a left-to-right orientation wherein *Knowledge Retention* and *Industry Taxonomy Standards* are considered early on, while *Technology Transfer Web Site* and *Core Utility Information Pilots* are addressed in the latter stages of the process. Projects which span more than one contribution area (Strategies, Standards, and Solutions) will contribute deliverables to the multiple areas as depicted.

The representation also shows a clustering for projects that are related and can provide synergies to one another. For example, *Organizational Development KM Strategies* and *KM Supporting Resources & Infrastructure* have a close affinity and ideally one could view them as a major set of initiatives executed in an overlapping manner. Similarly, *KM Core Planning Processes* are integral to the successful execution of both of the preceding projects. While the reality of the funding process and the number of competing research projects proposed each year dictate that these candidate projects be evaluated and approved for funding in a sequential manner, this clustering perspective underscores the necessity to treat these discrete projects as related and complementary elements of the industry's overall KM program.

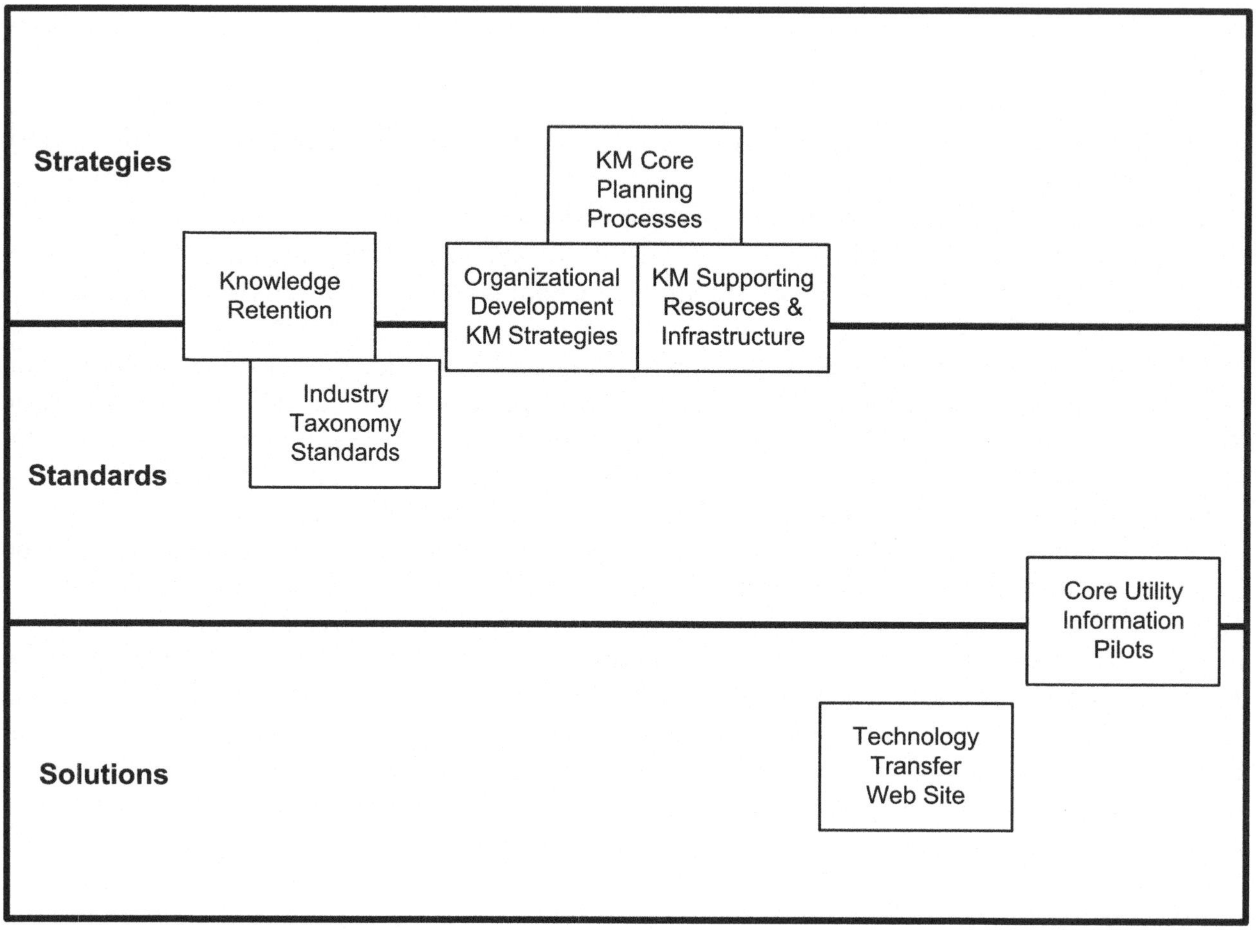

Figure ES.1 Candidate KM research project implementation sequence chart

Figure ES.2 shows the progressively embedded relationship among strategies, standards, and solutions as presented in this report. Ultimately, it is posited that the formulation of proper solutions rely upon defined standards which in turn are derived from the overall strategy.

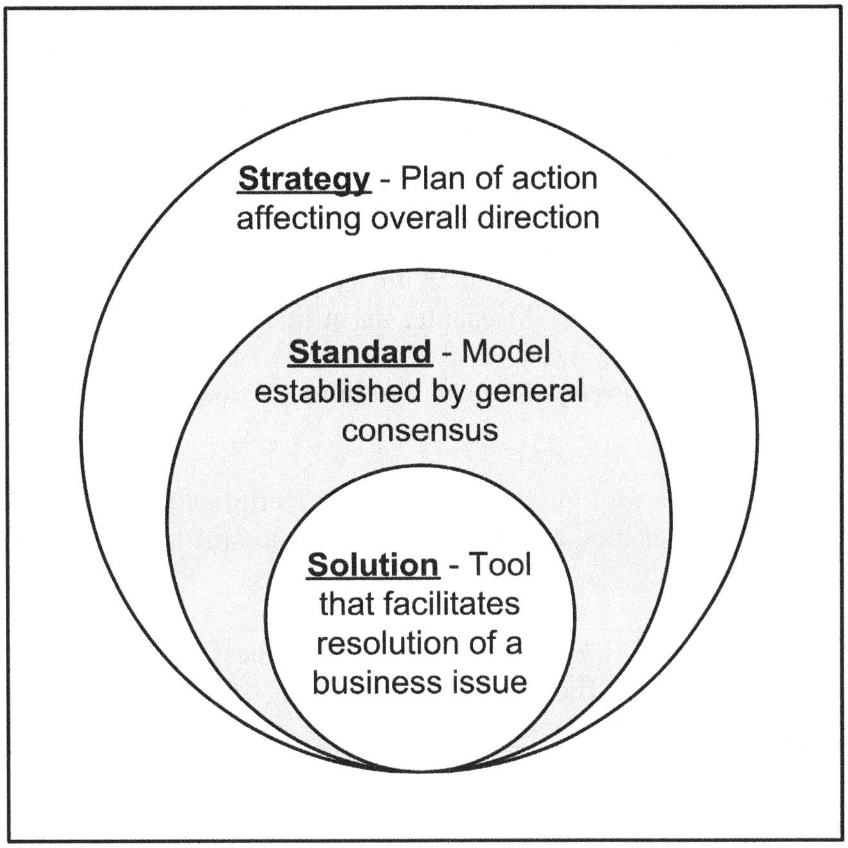

Figure ES.2 Candidate KM project contributions

Candidate Research Projects

1. Identify Strategies to Help Drinking Water Utilities Ensure Effective Retention of Knowledge

Objective: Identify strategies, tools, and techniques that utilities can use to retain knowledge from the retiring and shifting workforce. Identify key drivers, critical success factors, barriers, costs, and benefits related to implementing knowledge retention programs.

2. Define Industry Taxonomy Standards

Objective: Identify and develop an industry taxonomy and information architecture standard which can be leveraged as a KM-enabler for utilities.

3. Utility Organizational Development to Facilitate Knowledge Management Business Strategies

Objective: Develop a tool kit for drinking water utilities to evaluate and develop their organizational cultures so that they are supportive of successful implementation of knowledge management initiatives.

> Companies with tightly aligned technology with business strategy on average spend 17% less per user. (The Hackett Group, Study of 2,000 Global Organizations)

4. Best Practices for Aligning Core Planning Processes Necessary for Successfully Applying Knowledge Management

Objective: Investigate the relationship of a utility's core planning processes to implementation of a successful KM program. Identify how misalignment or an underdeveloped process impacts the successful deployment of a KM strategy and its resulting initiatives.

5. Identification of Required Supporting Resources and Infrastructure to Enable the Knowledge-Based Utility

Objective: Develop an enterprise integration model depicting an integrated flow (i.e., presentation, search, retrieval) of both structured (database) and unstructured (office applications) data and information contained in utility business systems and processes. Develop a supporting template providing utilities with the ability to readily understand the critical components (e.g., organizational skills/roles (technical and business), measurement approaches, readiness assessment, and security/business continuity/operations) that are necessary to support the implementation and maintenance of a KM strategy.

6. ***Application of Communities of Practice to the Drinking Water Industry***

Objective: Investigate the application of Communities of Practice (CoPs) to the drinking water community. Design, develop, and pilot a CoP for the drinking water industry and document findings relative to identified success criteria. Identify costs and benefits, key drivers, critical success factors, and barriers to successful implementation and industry adoption of CoPs.

7. ***Pilot Business-Enabling Core Utility Information Solutions***

Objective: Identify common industry and core utility processes (e.g., asset management, drawing management, O&M manuals, etc.) that can benefit from KM-based solution approaches. Investigate and document existing "best-in-class" approaches in the industry for these common processes and develop a comprehensive "best practice" model for common core information life cycle management approaches that can be leveraged across the industry by individual utilities.

CONCLUSIONS

This section contains a summary of the main conclusions and recommendations that were reached as a result of the project's research and investigation into the KM domain. In the body of the report, the topics addressed here are discussed in greater detail.

- The business drivers (e.g., growth, attrition, customer intimacy, productivity, reduced costs, improved asset management) that motivated other industries to undertake KM initiatives are very similar to the drivers and challenges that the drinking water industry faces.

- Applications of KM can benefit the drinking water utility sector much like other industries have benefited from them. Practical lessons can be learned from these other industries and these lessons can transfer over to the drinking water industry, resulting in a reduced learning curve for realizing benefits from applying KM strategies.

- The research identified that several of the participating utilities appear to be further along with not only having information in electronic form, but also with utilizing Web-based tools and technologies to facilitate access to collected information. These findings suggest that additional research within the utility sector investigating these exemplary implementations will be beneficial to further understanding the value proposition of KM to the industry.

- Successful application of KM embodies the critical elements of: people, culture, process, and technology. While each of these elements contributes to a KM initiative's success, it is the people element considered in context with the organization's culture that is most important to achieving positive and sustainable results.

- Utilities need not wait for future research results to begin fostering a learning culture and implementing organizational change agents that facilitate establishing a collaborative environment. This environment must be designed to promote learning, innovation, horizontal communication, empowerment, and teamwork through efficient processes based on standard and practical approaches. Ideas to consider in developing a learning culture include incorporating knowledge transfer objectives in employee reviews, holding lunch-and-learns where subject matter experts speak on a topic as a means to facilitate knowledge transfer, and establishing metrics and processes for recognizing individual or team contributions to increasing the organization's information and knowledge stores and meeting knowledge transfer objectives. Importantly, the Utility Manager will want to develop a clear message describing knowledge sharing as the preferred norm within the organization and incorporating techniques which promote and reinforce this behavior. (For examples of techniques currently being used to motivate personnel and improve knowledge transfer, please refer to Appendix B.)

- Various studies have detailed how organizational learning is not an easy goal to achieve and that organizations have a tendency to forget. A concept that must be considered in the context of KM and one closely aligned with existing research results is "organizational amnesia." (For more on organizational amnesia, please refer to Appendix C.)

- Unfortunately, technology is sometimes used interchangeably with the term "Knowledge Management." Although some early adopters thought implementing a database or other software tool constituted knowledge management, it is now known, through the documentation of case studies and academic research, that the role of technology is one of an "enabler." The Utility Manager must always remain cognizant that technology should never be the primary emphasis of a KM strategy.

- Although it is the project team's contention that industry decision makers should pursue incorporating KM strategies into strategic planning processes, the application of KM is not without challenges. When participants were asked to identify the main challenges that their utility would face in implementing KM within their organization, the hurdle most often cited dealt with available and appropriately skilled resources. In general, the respondents describe their organizations as spread thin by competing initiatives and struggling with maintaining and developing the proper mix of technical and project management skills. This observation underlines the importance for the internal KM champion to develop and present sound and pragmatic business cases that will substantiate the procurement of resources and skills necessary to support a KM initiative.

- In summary, the drinking water industry can benefit from implementing KM strategies. The drivers and challenges that influenced the adoption of KM as a management strategy in other industries are similar to what the Utility Manager is currently facing. Although implementation hurdles exist, they can be minimized by incorporating the following critical success factors in the KM strategy:

- The core processes of strategic planning, technology master planning, business/operational planning, capital improvement program (CIP) planning, and human resources (HR) life cycle activities must be viewed as intertwined and processes must exist to maintain their alignment. Misalignment of planning processes or the exclusion of a given core process from consideration commonly results in initiatives with significant cost overruns, projects that are abandoned mid-stream, and deliverables of the KM strategy that cannot be adequately maintained in production.
- Although the project participants identified lack of resources, shortage of appropriate skills, and competing initiatives as barriers to pursuing a KM pilot, it is important to realize that KM initiatives must not be thought of as independent initiatives but must be viewed as part of the utility's strategy and embedded as part of continuous improvement activities
- KM integration within the organization's overall strategic efforts does not minimize the importance of adequately and properly resourcing KM-related initiatives. Utility Managers looking to implement initiatives supporting KM must investigate the critical skills and resource requirements for all aspects of the initiatives' life cycle (e.g., planning, implementation, operations).

CHAPTER 1: PROJECT BACKGROUND

> Unlike capital, knowledge is most valuable when it is controlled and used by those on the front lines of the organization. (C.A. Bartlett, *Harvard Business Review*)

OVERVIEW

Knowledge Management (KM) is an important business strategy that has been embraced by many organizations, both nationally and internationally, across various industries. This trend differs from traditional management approaches in that it strives to create a learning environment in which personnel have on-demand access to an organization's data and information in such a manner that informed decision making and efficient and effective operations are a result. Key characteristics of this environment are that it is knowledge-enabled, promotes collaboration among stakeholders, and consciously incorporates people, processes, strategies, and technologies in the development and execution of strategic business plans.

The Awwa Research Foundation (AwwaRF) recognized the applicability of the business-oriented collaborative management principles inherent in KM to the water industry in order to enable more effective use of organizational data and information. Toward this end, AwwaRF sponsored a research initiative that started in the first quarter of 2004 with an objective of developing a comprehensive, prioritized list of potential KM research projects that pertain to the drinking water utility business. The primary goal of these candidate research projects is to enhance the creation, discovery, collection, organization, distribution, accessibility, and effective usage of relevant and current knowledge at all levels within a drinking water utility. These candidate projects would not only identify techniques and tools to assist with the creation and management of new knowledge but also facilitate leveraging knowledge already captured.

This research initiative's impetus is centered on the need for the industry to continue to move toward an environment that utilizes more business-oriented approaches. Both AwwaRF and Utility Managers realize that to be successful in the current challenging business climate, improvements must be made in the way that the key stakeholders (e.g., personnel, partners, boards, customers) interact and communicate. Additionally, the immense amount of data and information (Please refer to Appendix D for a discussion on the relationship of data, information, and knowledge) that the utility has created and stored over its life has not been used as effectively as possible in support of decision-making processes. Compounding this inefficiency is that the growth rate of this information mountain continues to accelerate. Drivers such as these justified an initiative that would take a close look at the emerging management strategy of KM and identify its potential and applicability for assisting the industry in overcoming these and other challenges.

The research team selected to lead this initial research project crafted a unique approach encompassing industry participation, bibliographic search, collaborative workshops, and outside-industry subject matter experts and practitioners. In formulating the proposed project approach, the research team constructed a model that defined a set of critical outputs that will further develop KM as a mainstream management strategy for the industry. These critical outputs centered around three key areas: Strategies, Standards and Solutions. The research team believes that the benefits that the industry is to realize from KM will be directly correlated to the degree that the industry develops and uses common approaches for solutions. Figure 1.1 illustrates the process model used to complete this project.

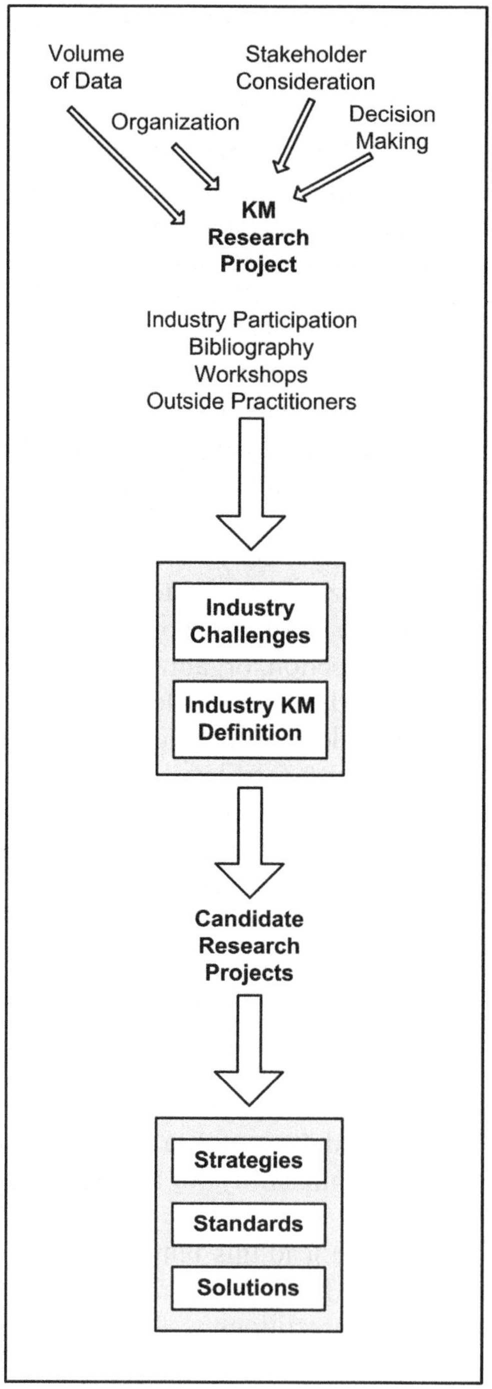

Figure 1.1 KM project model

Starting at the top and flowing downward, the model depicts how – through an approach of collaborative workshops made up of both project industry executives and nonproject industry executives (presentation and workshop at AMWA Annual Meeting, October 2004) – the industry challenges were captured, prioritized, and validated. The process mapped the challenges back to the project-initiated industry definition, confirming the applicability of KM for the industry. The challenges and definition were then used to identify potential candidate research projects to further define KM as a mainstream management strategy for the industry. All of the candidate research projects were constructed and targeted to support understanding in one or more of the three areas: Strategies, Standards and Solutions. It is through these tools that the industry will come to effectively and uniformly realize the benefits of incorporating KM into its management philosophy.

Through the process of executing this initial research project and identifying additional candidate research projects, the research team successfully achieved its goals of understanding KM, confirming its applicability to the industry, and recommending a slate of follow-up research projects to define a new management strategy to benefit not only individual utilities, but also the industry as a whole. Outputs from the candidate research projects will provide the industry with a combination of KM-based strategies, standards, and solutions that may be leveraged by Utility Managers to meet current and future challenges related to the acquisition, management, dissemination, and reuse of knowledge.

CHAPTER 2: RESEARCH APPROACH

OVERALL APPROACH

Specifically, the approach to this project consisted of performing the following key tasks (Figure 2.1):

1. Recruit industry participation
2. Research and implement a maintainable KM literature knowledge base
3. Conduct workshops and surveys
4. Specify and deliver a set of outcomes
5. Foster ongoing communications.

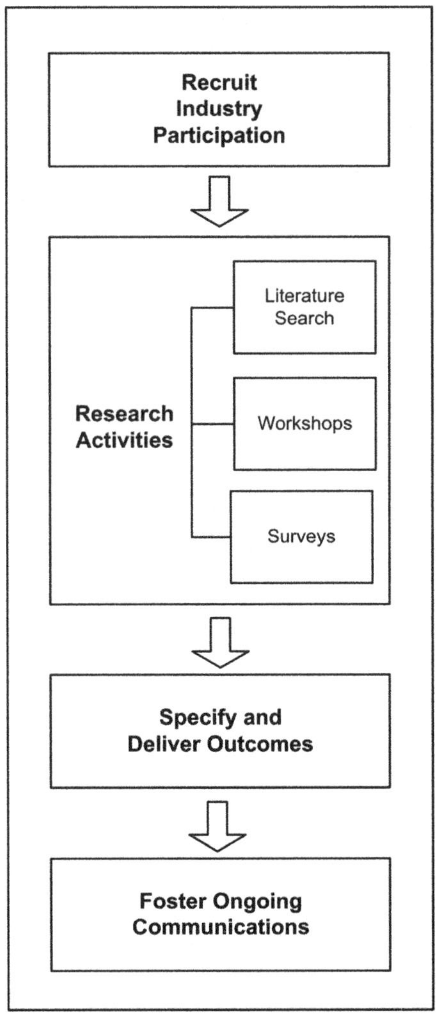

Figure 2.1 Research approach

RECRUIT INDUSTRY PARTICIPATION

The research team viewed this project as the beginning step toward understanding the knowledge management domain and investigating KM's potential for becoming a part of the industry leadership's management strategies. As such, cross-sectional representation of industry personnel was very important to obtaining the proper perspective of KM's application to usefulness for solving problems and overcoming challenges that the industry faces. The research team's ultimate goal for this initial group of participants is for them to develop and evolve into a community of practice (CoP) within the industry with a primary role of increasing the awareness relative to KM's benefits and best practices. In this context the research team defines CoP as a group of people who share a concern, mandate, or sense of purpose. Several studies of KM initiatives in other industries have found CoPs to be an important ingredient to their success. These studies identified CoPs as functioning as boundary-spanning teams in organizations, serving as catalysts for discovering and sharing best practices, and facilitating cross-functional communications. Over time it is hoped that this CoP will mature and be leveraged by the Foundation to continue to internalize the applicable benefits of KM strategies and become a means for institutionalizing and mainstreaming these strategies within the industry.

The research team, in conjunction with guidance from the Project Advisory Committee (PAC), consciously strived to recruit utilities that represented the diverse characteristics of the industry. To that end it was very important to pull together a team from both the national and international community representing diverse customer bases and various geographical regions. Table 2.1 defines the project's participants.

Table 2.1 AwwaRF knowledge management project participants

Utility	Primary Participant Name
Anchorage Water and Wastewater Utility (Alaska)	Mark C. Premo - General Manager
Austin Water Utility (Texas)	Brownlee Bowmer - Chief Information Officer
Chicago Department of Water Management (Illinois)	John Spatz - Deputy Commissioner
City of Cleveland Division of Water (Ohio)	Ron Petrich - Program Manager
Columbus Water Works (Georgia)	Cliff Arnett - Senior Vice President - Operations
Dallas Water Utilities Department (Texas)	Robert M. Johnson, P.E. - Director
Greater Cincinnati Water Works (Ohio)	Steven Hellman - Business Service Superintendent
Detroit Water and Sewerage Department (Michigan)	Kevin Quiggle - Information Services Manager
EPCOR Water Services (Alberta)	Audrey A. Cudrak, M. Eng., P. Eng. – Manager, Process Services (PAC Member)
County of Henrico Department of Public Utilities (Virginia)	Arthur D. Petrini, P.E. - Director of Public Utilities
Lafayette Utilities System (Louisiana)	Don Broussard - Water Operations Manager
Las Vegas Valley Water District (Nevada)	James W. Ellisor - Director, Information Systems
Minneapolis Water Works (Minnesota)	Adam J. Kramer, P.E. - Director of Water
Oakland County Drain Commission (Michigan)	Kevin R. Larsen - Chief Deputy Drain Commissioner
Portland Water District (Maine)	Gordon Johnson - Water Operations Engineer
Portland Water Bureau (Oregon)	Mark Knudson - Director, Operations & Maintenance
San Diego County Water Authority (California)	Marilyn Young - Administrative Services Department Manager
Seattle Public Utilities (Washington)	Dave Muto – Manager of Water Supply & Treatment (PAC Member)
Thames Water Utilities (United Kingdom)	Linda Godfrey - Knowledge and Business Services Manager (PAC Member)
APQC	Jim Lee - Senior Strategy Advisor (PAC Member)

This participant list represents a solid cross-section of the industry's demographics as demonstrated by the following tables (Tables 2.2, 2.3, 2.4, and 2.5).

Table 2.2 Average number of employees (19 total responses)

Number of Employees	Number of Responses
<100	0
100-249	4
250-499	5
500-749	2
750-1000	0
>1000	8

Table 2.3 Average number of residential water customers (population) (18 total responses)

Number of Customers	Number of Responses
<75K	0
75K - 249K	6
250K - 499K	0
500K - 999K	3
1M - 1.99M	5
≥2M	4

Table 2.4 Average water treatment capacity (17 total responses)

Capacity	Number of Responses
<10 MGD	1
10-24 MGD	0
25-49 MGD	4
50-99 MGD	3
100-249 MGD	3
250-499 MGD	3
500-749 MGD	1
750-999 MGD	0
>1 BGD	2

Table 2.5 Service area size in square miles (15 total responses)

Square Miles	Number of Responses
<100	0
100-199	5
200-299	3
300-399	0
400-499	0
500-599	0
600-699	3
700-799	0
800-899	1
900-999	0
>1000	3

RESEARCH AND IMPLEMENT A MAINTAINABLE KM LITERATURE KNOWLEDGE BASE (BIBLIOGRAPHY)

To cultivate an environment conducive to reaching the goals of the project, it was important to develop a baseline of understanding within the team related to terms, definitions, fundamental principles, common assumptions, and expectations of KM. Additionally, researching and discussing case studies pertaining to real-life implementations was extremely important to reaching a solid basis in support of achieving the overall vision of the project. These objectives caused the research team to investigate and develop an applicable library of readings discussing the following subjects: KM primers, cultural considerations, best practices, and industry-specific case studies. The research team looked to executive-focused correspondence for much of this material with the resulting list representing well-known KM subject matter experts and practitioners from periodicals such as *Harvard Business Review, CIO, Darwin, Chief Executive, Computerworld,* and *KMWorld.* A wide range of professional conference proceedings, engineering informational databases, and other professional journals and publications was also consulted.

Critical to this effort was designing a mechanism for easily sharing and collaborating on these readings and, most importantly, putting in place a tool to facilitate communicating, sustaining and maintaining this valuable collection of knowledge. The research team developed a Web site utilizing a low-cost hosted portal solution to achieve this objective. This portal solution provided various functionalities to support and nurture team-based collaborative efforts. This solution provided a means for project participants to come together as a community and continue to evolve KM as a key strategy and management discipline for the industry. Importantly, it established a mechanism for the industry and AwwaRF to continue collaborative work with respect to KM, and provided a central source for knowledge, best practices, lessons learned, and CoPs.

CONDUCT WORKSHOPS AND SURVEYS

To facilitate collaboration among the project participants and to provide a forum for discussion of pertinent readings and case studies, four workshops were facilitated by the research team. Three of these workshops were virtual in nature, carried out through the use of a Web-based collaboration tool complemented with voice conferencing. These workshops occurred in 2004 and early 2005, and covered specific themes and provided a mechanism for the diverse participants to respond to questions and observations from their peers. Also discussed during these sessions were the results of several surveys that the research team developed as a means for gathering specific data points, along with qualitative-based insight from each utility into key elements of KM. Each survey consisted of approximately 15-20 questions centered on several topics consistent with KM considerations. The major topics covered were:

- Participant Expectations from Project Participation
- Utility Demographics and Characteristics
- Use of Metrics to Guide Utility Operations and Assess Performance
- Insight into the Organization's Culture
- Planned Initiatives in Support of KM as Described in the Draft Industry Definition
- KM-Enabling Technologies Currently in Use or Being Planned
- Observations of Potential Hurdles to Implementing a KM Pilot.

The day before the start of the June 2004 annual AWWA conference in Orlando, the participants took part in an on-site project workshop facilitated by the research team. During this workshop several team members presented examples of initiatives being carried out at their utilities that addressed critical business issues. To provide a perspective from outside the industry, representatives from two organizations prominent in the study and application of KM – the National Aeronautics and Space Administration (NASA) and APQC – were asked to present their experiences with KM. These presentations proved quite beneficial in that they highlighted practical applications of KM and provided insight into how other practitioners are addressing common business challenges through the application of KM strategies. Also at this on-site meeting, the research team facilitated a brainstorming session during which the team members developed ideas for utilizing KM strategies to assist with solving industry problems. These ideas served as the genesis for the additional KM-centered candidate research projects proposed by the project team.

SPECIFY AND DELIVER A SET OF OUTCOMES

To provide focus and achieve the overall goals of the project, the research team developed a road map that articulated a specific set of project outcomes. This set includes the following deliverables:

- Functioning KM CoP
- KM Stewards and Champions
- Development of an Initial KM Body of Knowledge
- Candidate Industry Process Areas for KM Applicability
- Industry Direction on Candidate Projects to Include Prioritization.

Several of these deliverables, such as KM CoP and KM Stewardship, have had a foundation established by virtue of the collaboration facilitated by the project Web site and virtual and on-site working sessions. As the industry and Foundation continue to research KM strategies, the institutionalization of these less-tangible deliverables will be fundamental to embedding KM best practices within the core competencies of the industry.

The candidate process areas and corresponding potential KM research projects have vast practical benefits that can be realized by the industry. Successfully implementing these candidate projects will benefit the industry through increased employee productivity, improved decision-making ability, reduced costs, and increased responsiveness to stakeholder demands and needs.

FOSTER ONGOING COMMUNICATIONS

Although the KM research project provides the industry and Foundation valuable insight into the critical elements necessary for successful implementations of KM and its applicability to the industry, it is only the beginning. It is extremely important that the industry maintains the momentum that the project has initiated and continues to promote education and awareness for industry decision makers pertaining to the evolution and maturing of KM as a management strategy. During the course of the project, the research team facilitated interactions, awareness, and collaboration by leveraging the project Web site as the central focus for communications, meeting minutes, presentations, and pertinent readings. The body of knowledge that the site contains can continue to be valuable by enhancing it with the publication of lessons learned and success stories that future industry KM initiatives will bear. To realize the benefits of KM-based management strategies, a conscious effort must be made to facilitate and nurture the ongoing interaction of the embryonic KM community within the industry. Also, as the overall KM body of knowledge continues to grow, the current bibliographic database should be maintained as it represents a vast amount of information and knowledge pertaining to the domain of KM. One measure of success relative to KM's application to the industry will be the number of industry-specific case studies and articles that this database contains in the future.

Another important aspect of the communications strategy is the ongoing evangelizing and awareness presentations that the research team performed during the course of the project. The research team, with the support of several of the team members, made presentations of the objectives, survey results, and project status at several national and local industry conferences and workshops. This activity was well-received and generated significant interest from the various audiences. For the researchers, it confirmed that KM strategies have applicability in the industry and there are utility leaders and organizations that are quite interested in learning more about the practicality of KM, and its potential for assisting them in solving business problems. The presentations performed during the project are summarized in Table 2.6.

Table 2.6 KM project presentations

Conference Name	Location	Date
AMWA Annual	Florida	October 2004
Louisiana Water Quality Technology Conference	Louisiana	December 2004
Joint Management Conference (AWWA-WEF)	Georgia	March 2005
California Nevada Section AWWA	California	April 2005
AwwaRF Workshop	Ohio	April 2005
Central States Water Environment Association	Minnesota	May 2005
AWWA Annual	California	June 2005
Michigan Water Environment Association	Michigan	June 2005
Nuclear Information and Records Management Association	Florida	August 2005
Rocky Mountain Section AWWA	New Mexico	September 2005
AwwaRF Technology Transfer Conference	Missouri	September 2005
AwwaRF Technology Transfer Conference	Michigan	October 2005
The National Forum for Black Public Administrators*	Texas	April 2006
California Water Environment Association*	California	April 2006

*Submitted abstract

CHAPTER 3: FINDINGS

> Many modern information systems dysfunctionally add to organizations' inertia. Access to more information and more advanced decision aids does not necessarily make decision makers better informed or more able to decide. (B. Hedberg and S. Johnsson, *Accounting, Organizations, and Society*)

SURVEY OVERVIEW

An important tool utilized during the research was Web-based surveys (see Appendix A), completed by project team members, that covered key topics pertaining to industry demographics, characteristics, trends, challenges, culture, performance measures, and technology. As appropriate, the research team followed up with the survey respondents to obtain clarification or additional input. The surveys were structured such that each one focused on several key topics and collected both quantitative and qualitative results from the participating utilities describing the landscape and environment to which KM applications will be applied. As needed, the primary project participants solicited input from their utility personnel for completeness and consensus.

The surveys were meant to engage the participants and elicit general information that could be used as a baseline of the participant and their respective utility characteristics. The surveys were not meant to gather data for the purpose of presenting a statistical comparative analysis.

Survey #1 Objectives

The first survey obtained data and information pertaining to the demographics of the project participants and characteristics of their respective utilities. Respondents were also asked to describe their personal understanding of KM and any prior experience with KM initiatives. Feedback pertaining to both personal and organizational expectations regarding project participation was solicited. All 19 utility participants responded to Survey #1.

Section One: Respondent Demographics

Job Level – Participant respondents ranged from Executive Director to Manager. The majority of participants have achieved the job level of Assistant Director.

Functional Area – Most participants were affiliated with Executive Team, Information Systems, and Operations. Additionally, participants indicated being associated with the functions of Engineering, Asset Management, Finance/Accounting, and Procurement.

Years in Position – The average participant had 4-10 years of experience in his/her current position. Several participants indicated tenures of 11-20 years, with a maximum of one 20-year response. Conversely, one participant had held the position less than 1 year.

Years with Organization – The average response indicated a tenure of 10-20 years with the utility. The maximum tenure range was 20+ years of experience, while the minimum tenure range was 1-3 years.

Annual Conference – Almost universally, participants indicated participation in the AWWA annual and sectional conferences. Several respondents indicated regular attendance of the annual Water Environment Federation (WEF) conference.

Section Two: Utility Characteristics

Public or Private – The majority of participants represented utilities that are publicly held and operated; however, privately held utilities were also represented.

Service Offered – All participants offered Retail and Wholesale Water Services while a majority also offered Wastewater Treatment and Collections Services. Notably, Solid Waste, Electrical, Billing, and even Telecommunications are offered or are being considered by some of the participants.

Number of Employees – Participating utilities had an average of 500-750 employees. Several participants employed more than 1,000 personnel, while the smallest participants employed 100-250 personnel.

Number of Residential Water Customers (i.e., Total Population) – Respondents indicated an average range of 500,000-1,000,000 customers. Several participants indicated customer bases in excess of 2,000,000, while nearly half of the respondents indicated customer bases of less than 250,000.

Number of Wastewater Customers (i.e., Total Population) – Participants offering Wastewater Services indicated on average 250,000-500,000 customers. The maximum exceeded 2,000,000 customers. The minimum served was fewer than 75,000 customers.

Average Water Treatment Capacity – Respondents indicated an average capacity of 50-99 millions of gallons per day (MGD). Larger respondents indicated capacities in excess of 1 billion gallons per day (BGD), while smaller utilities indicated a capacity of less than 10 MGD.

Peak Water Treatment Capacity – Respondents indicated a peak capacity of 100-249 MGD per day. Larger respondents indicated capacities in excess of 1 BGD, while smaller utilities indicated a capacity of less than 10 MGD.

Average Wastewater Treatment Capacity – Respondents indicated an average capacity of 100-249 MGD per day. Larger respondents indicated capacities in excess of 1 BGD, while smaller utilities indicated a capacity of less than 10 MGD.

Peak Wastewater Treatment Capacity – Respondents indicated a peak capacity 250-499 MGD per day. Larger respondents indicated capacities in excess of 1 BGD, while smaller utilities indicated a capacity of less than 10 MGD.

Service Area – The average service area size of participating utilities was 339 square miles. Several utilities indicated service areas of 750-1,000 square miles. Smaller utilities reported service areas of 100-200 square miles.

Annual Operations Budget – The average annual operating budget of participating utilities was $100-$150 million. On the high end, a number of participants indicated budgets in excess of $200 million. On the low end, several utilities indicated budgets of $10-50 million.

Section Three: Key Organizational Issues and Pressures

Key Organizational Issues and Pressures – Respondents were asked to consider a list of common industry issues and pressures based on the research team's knowledge of external influences and industry trends. This question generated significant interaction and discussion by the participants. When asked to rank and prioritize the top three issues and pressures facing their utility, most respondents struggled to identify a clear-cut answer. Interestingly, after the discussions, most utilities settled on the following key areas (listed in order of importance):

- Aging Infrastructure (clearly the #1 issue and challenge cited by the majority of participants in their top 3 challenges)
- Rate Pressures and Security (only one vote separated these two issues)
- Workforce Recruitment, Attrition, Retention and Succession Planning (despite being expressed as three separate items, these categories were ultimately combined as a group by the research team, based on the fact that most participants believed the categories represented a set of related issues).

Other issues and pressures receiving votes were:

- Water Resources (that is, finding or sustaining high-quality water resources, often to meet the challenges of growth)
- Expansion of Customer Base (that is, the challenge of meeting the needs of a growing customer base)
- Regionalization
- Workforce Motivation (that is, continue engagement and appropriate focus on further advancing improvements already obtained)
- Politics and Regulatory Issues.

Tools Currently Used – Respondents were asked to review and indicate from a list of technologies those that were used by themselves or the utility in performing daily job functions. The research team's goal with this question was not to identify which technologies were being used, but instead which technologies were not being employed by the respondents. The top three technologies not employed by participants were:

- e-Learning Software
- Web-based Discussion Forums
- e-Meetings.

This list did not surprise the research team as these technologies are still in the early stages of adoption globally by organizations (private or public). Surprisingly, a fair number of participants indicated that Project Management and Laboratory Information Management System (LIMS) applications were not deployed or used on a regular basis within their organizations.

Section Four: Participants' Personal Experiences and Expectations

Describe What You Believe Knowledge Management (KM) Means – Most respondents commented in terms of being able to capture information from the current workforce and store it in such a way that it can be disseminated and reused on an as-needed basis by the future workforce. Other concepts mentioned included:

- Creation and enhancement of a corporate history repository, and
- Treating information as a corporate asset.

Prior Experience with KM – Respondents were asked to classify their previous experience with or exposure to KM. More than 50% of respondents indicated that they had no previous experience. Approximately 40% indicated their only previous experience came from reading articles on the subject. Only 10% indicated having participated on a KM-related project.

Motivation to Participate in This KM Research Project – Respondents were asked to classify their motivation to participate from a list of possible motivators:

- I understand what KM is and believe a KM strategy can benefit our utility
- I understand what KM is and would like to learn how KM can benefit our utility
- I want to understand how other organizations have used KM to solve problems
- I know little about KM and want to learn more about it
- I know my organization does not effectively manage information and I think KM may act as an enabler to improve collaboration within teams and reuse of information

Respondents were allowed to check as many motivators as warranted for this question. By far, the two leading motivators for participants were:

- I want to understand how other organizations have used KM to solve problems
- I know my organization does not effectively manage information and I think KM may act as an enabler to improve collaboration within teams and reuse of information.

Expectations from Personal and Organizational Perspectives – Common themes were expressed by participants in answering these questions:

- Personally – Participants indicated they wanted to understand what KM is and how it could be applied to the Water Industry.
- Organizationally – Participants indicated they hoped to be able to communicate what KM is and how it can be applied for benefit within their utility.

> The computer is merely a tool in the process… To put it in editorial terms, knowing how a typewriter works does not make you a writer. Now that knowledge is taking the place of capital as the driving force in organizations worldwide, it is all too easy to confuse data with knowledge and information technology with information. (P.F. Drucker, *Managing in a Time of Great Change*)

Survey #2 Objectives

The second survey's objective was to understand each utility better from the following three dimensions:

- Metrics and measures used to manage the utility
- Cultural characteristics
- The utility's focus for Capital Improvement Program (CIP) initiatives and their approach and process for managing CIP initiatives.

Seventeen of the 19 utility participants responded to Survey #2.

Section One: Business Metrics

Key Measurements Used – All of the respondents except one identified multiple measures that are being used as key performance indicators. As might be expected, different measures are used across the participating utilities with differing degrees of importance being associated to a given measure, depending on the utility. Although the individual measures varied, they consistently were reflective of several core dimensions: customer, operations, employee, financial, and compliance. Many of the respondents talked about a hierarchical structure to measurement in that different measures were captured and reviewed at the different levels of the utility's organizational structure. Although a few of the respondents described a very structured reporting and analysis process of the measurements relative to plan or target, the majority discussed a more informal and exception-based review and analysis process. Several of the respondents commented that some of the metrics being maintained are questionable in their value due to suspect accuracy, lack of standard definitions, and inconsistency of underlying algorithms. Unanimously, the respondents indicated that this area is one of increased focus.

Familiarity and Participation in Benchmarking Activities – The survey contained several questions that asked the respondents to describe their familiarity with the term "benchmarking" and whether their utility performed or participated in any benchmarking activities. Additionally, the respondents were asked to describe whether the benchmarking process was formal and to comment on the frequency of data collection.

All of the respondents were familiar with the term "benchmarking," and 14 of the 17 respondents stated that their utility participated in specific benchmarking activities. Of the 14 respondents, 8 described their benchmarking process as a formal activity. Several of these respondents indicated that their benchmarking activities took place more than 3 years ago. Data collection periods varied from monthly to quarterly, depending on the measure, with the frequency of review and analysis mirroring the data collection periods.

QualServe Initiative – When asked to comment on their familiarity with the QualServe initiative, 13 of the 17 respondents were aware of it but many commented that they would like additional understanding of this program. Several of the utilities use it as a cross-check for some of the metrics that they collect and analyze, but no one spoke of using QualServe as a primary means to manage their day-to-day business.

> 80% of all organizational knowledge is not centralized. It is found on desktops, hard drives, and stuck in drawers. (*KMWorld*)

Section Two: Organization Culture

Sharing, Collaborating and Learning Environment – Several questions were asked of the respondents to gauge cultural and organizational behaviors with respect to sharing information among peers, supporting processes (both formal and informal) that promote collaborating across functional units and teams, and how failure was perceived within the organization.

Universally, the respondents answered that sharing and collaborating was promoted and encouraged within the organization, although most respondents qualified this response by stating that there were still significant challenges, both people and culturally related, that needed to be overcome. Consequently, most viewed their utility as just beginning to realize the benefits that these approaches and methods have on improving the business operations. Additionally, the majority of respondents described their environment as one that viewed failure as an opportunity to learn, although this, too, is still not institutionalized.

Supporting Policies, Procedures and Processes – The survey asked the respondents to comment on the existence of formalized methods, policies and processes that are followed to capture information, lessons learned and best-practice data so that it can be reused and leveraged over time in support of solving business issues. While most of the respondents' organizations have made minimal progress in this area, it is noteworthy to mention that, within the project participants, several utilities (e.g., Thames Water Utilities, Greater Cincinnati Water Works, Columbus Water Works, Anchorage Water and Wastewater Utility) are further ahead with established and formalized processes, procedures and mechanisms for capturing and addressing lessons learned, identifying best practices, and harvesting expertise from subject matter experts. There appears to be an opportunity for these leading organizations to be leveraged by future AwwaRF activities for the purpose of establishing standards, methods, lessons learned, and guidelines that other utilities can utilize.

Section Three: Planned Initiatives

Capital Budget and Initiative Focus – Respondents were asked to comment on the size of their capital budget and describe its area of emphasis. The range for capital budget size was $10 million to $800 million annually. Predictably, the size of the capital budget tracked closely with service area and customer base. Notably, and as expected, the vast majority of respondents were able to answer this question easily, indicating that this information is freely communicated throughout the organization.

The respondents described several common initiative focus areas. The areas discussed were closely aligned with the responses collected in survey #1 which described the key business drivers. The following summarizes the common themes:

- Infrastructure – New and Maintenance
- Compliance – U.S. Environmental Protection Agency (EPA)
- Operating Efficiencies – Automatic Meter Reading (AMR)
- Customer – Billing Systems.

Standards, Sound Techniques and Project Management – The final questions of the survey asked respondents to evaluate and comment on their organizations' process for nominating, reviewing, approving and implementing capital projects. They were also asked to describe the degree of standardization and common methods that are used to execute these initiatives.

Generally the respondents felt that there were solid and effective processes in place for nominating and approving CIP initiatives. Most organizations use a team approach with well-identified and established selection criteria. In the area of project review, the responses were generally positive but more varied with respect to consistency and frequency of the review process.

Likewise when the respondents were asked whether their organizations followed established standards and applied a disciplined approach to executing CIP initiatives, their responses were consistent. Although there were examples of this type of sound project management being used, they were not used consistently on every project. The general response was that although improvement with respect to structure, discipline, and project management techniques was happening, it is an area in which most of the respondents felt they could continue to improve.

> Percentage of employees who rely upon Technology to do their job rose by 45% between 1999 and 2002. (The Hackett Group, Study of 2,000 Global Organizations)

Survey #3 Objectives

The third survey conducted by the research team focused on identifying what knowledge management-enabling technologies currently exist or are planned for implementation within the project's participating utilities. Additionally, the survey collected feedback and insight into potential hurdles with respect to implementing a knowledge management pilot. Thirteen of the 19 utility participants responded to Survey #3.

Section One: Enabling Technologies

Database Software, Report Generation, and Business Intelligence Software – All of the respondents identified having currently deployed enterprise-level (e.g., Oracle, MS SQL) database solutions. Most of the participants also stated that report generation and business intelligence software exists within the environment in the form of a Business Objects product suite (e.g., Crystal Reports). Interestingly, there is planned growth with respect to these tools, as approximately 60% of the respondents identified 2005 planned deployments in addition to the currently implemented configurations.

Document Management, Content Management, and Imaging Software – Ten of the 13 respondents noted that software tools supporting the electronic capture and management of documents and drawings currently exist or are being planned for deployment in 2005.

Collaborative Software – Forty-six percent of the respondents currently use collaborative tools such as video conferencing, NetMeeting, and WebEx in support of key business processes. Two respondents identified that they plan to deploy collaborative tools in 2005.

Web Environment – To gauge the utility's Web computing environment for readiness with respect to distributing, presenting, and collaborating on information and knowledge, the respondents were asked to describe their Web environment architecture and its usage by the different functional groups within the utility.

Seventy-seven percent of the respondents described having a Web-based environment capable of delivering applications and information such as reports, customer service frequently asked questions, customer billing information, and water quality information. Although most of the respondents described an environment that is still evolving, several appear to have a highly integrated environment that has many of the business systems accessible through an intranet. Along with this integrated infrastructure, these "best-in-class" organizations also spoke of information architectures, processes, procedures, and security policies for managing information access, delivery, presentation, and security. It appears to the research team that further investigation of these leading organizations could provide benefits to other utilities as they continue to build out their enabling knowledge management infrastructure.

> There is a huge impedance mismatch between the unstructured world of the fax, the phone call, e-mail, and dealings between people, and the structured world of these back-end IT systems. We have to fix that. (Bill Gates, *Business 2.0*)

Section Two: Information in Electronic Format

Digitized Information Assets – The respondents were asked to identify what types of information the utility currently has in electronic format and to highlight what information is planned for conversion to electronic format in 2005. Not surprisingly, all of the respondents listed a fair amount of high-usage information such as engineering drawings, engineering standards, operations and maintenance (O&M) manuals, permits, and maps as currently being in electronic format. Most of these items were also listed as being planned for conversion to electronic format. To provide perspective on the types of information in electronic form, the respondents were asked to select a percentage range of how much of this information was paper-based versus in electronic form. Approximately 50% of the respondents believed that they had between 34% and 50% of their information assets in electronic form.

Tables 3.1 and 3.2 illustrate the respondents' replies using the categories presented in the survey.

Table 3.1 Percentage of information assets in electronic form (13 total responses)

Percentage	Number of Responses
0% - 25%	2
26% - 33%	1
34% - 50%	6
51% - 75%	3
>75%	1

Table 3.2 Information assets in electronic form

Information Asset Type in Electronic Format	Number of Responses
Engineering Drawings	12
Water Line Maps	12
Engineering Standards	10
O&M Manuals	10
Sewer Line Maps	10
Permits	9
Contracts	7

Section Three: Planning Process

Information System/Information Technology (IS/IT) Master Plan – Several questions were asked of the respondents regarding the IS/IT master planning process and this planning activity's alignment with the overall business planning function of the utility.

Approximately 70% of the respondents stated that there currently exists an IT master plan that includes processes for updating these plans. These update processes typically take place every 2-3 years. Of the 4 respondents reporting no current IT master plan, 3 state that plans exist to create one. Interestingly, of the 70% reporting that an IS/IT master plan exists, only 40% described an integrated process with the business planning efforts.

Section Four: Challenges

Knowledge Management Pilot – The survey's last question asked the respondents to describe the challenges and hurdles that need to be overcome prior to their utility implementing a knowledge management pilot. Table 3.3 lists the main issues that the respondents described. As the data in the table shows, the majority of respondents believed that technology was not the main hurdle, but resource constraints and appropriate skills provided the biggest challenge.

Table 3.3 Identified challenges for implementing a KM pilot

Identified Challenge	Number of Responses
Resource Constraints and Proper Skills	8
Adequate Budget	3
Data and Hard Copy Conversions	3
Information Design, Information Architecture	2
Information Standards	2
Competing Initiatives	2
Management Consensus	1
Adequate Infrastructure	1

CHAPTER 4: CONCLUSIONS

ESTABLISH A KM DEFINITION FOR THE WATER INDUSTRY

Early on, it was recognized that the development of a working definition for KM was a critical step in uniting the project participants around a common purpose. Reaching this consensus established the beginnings of a paradigm which can be further evolved by critical-thinking leaders and strategists within the industry. The current working definition combines themes found in definitions of KM practitioners from outside the industry with industry-specific criteria. As a result of the project team's increased insight and awareness of the KM domain, the definition was modified several times through the course of the project and is currently described as follows:

KM is a business strategy by which a water utility consciously identifies, captures, indexes, manages, and stores experiences, data, and information and provides methods for easily accessing and acting upon these collective assets in a collaborative environment optimizing the use of people, processes, and technology in support of:

- *Effective decision making*
- *Assuring compliance*
- *Improving performance*
- *Innovation*
- *Business continuity, all on a timely and sustainable basis.*

Although most project participants represented public utilities, private utilities were also represented. To comprehend the business model for private organizations, a modified definition is suggested which would include the addition of a new final sentence:

KM is a business strategy by which a water utility consciously identifies, captures, indexes, manages, and stores experiences, data, and information and provides methods for easily accessing and acting upon these collective assets in a collaborative environment optimizing the use of people, processes, and technology in support of:

- *Effective decision making*
- *Assuring compliance*
- *Improving performance*
- *Innovation*
- *Business continuity, all on a timely and sustainable basis.*

KM embodies processes by which the organization creates wealth (i.e., profit) by leveraging its intellectual and knowledge-based assets.

INDUSTRY CHALLENGES ARE NOT UNIQUE

The business drivers that motivated other industries to undertake KM initiatives are very similar to the drivers and challenges that the drinking water industry faces. Importantly, these challenges are also common across the project's participant utilities regardless of size, geography, or customer base. During the course of the project, the participants were asked to identify and prioritize the main pressures and challenges currently facing their organizations. This list is summarized by priority in Table 4.1.

Critical pressures such as outdated infrastructure, rate constraints, aging workforce, workforce retention and recruitment, increased security, and audit and reporting requirements are not unique to the industry. For example, the impending workforce change due to the baby boomers reaching retirement age is something that all organizations, both private and public, are facing. Likewise, aging infrastructure assets and increased regulations and reporting requirements have implications for government agencies such as NASA as well as industries like energy and automotive.

Applications of KM can benefit the drinking water utility sector much like other industries have benefited from them. We believe there are practical lessons to be learned from these other industries and assert these lessons can transfer over to the drinking water industry, resulting in a reduced learning curve for realizing benefits from applying KM strategies. Importantly, the research identified that several of the participating utilities appear to be further along with having information in electronic form and utilizing Web-based tools and technologies. These leading organizations are leveraging these tools and incorporating people-centered processes to capture, manage, and integrate information from key business systems while storing it in a form that facilitates its access and usage. Additionally, formal processes and settings supporting communication, collaboration, and knowledge sharing are being incorporated into the operational methods within their utilities. These findings suggest that additional research within the utility sector investigating these exemplary implementations will be beneficial to further understanding the value proposition of KM to the industry.

Table 4.1 Most pressing challenges for the industry

Type of Challenge
Aging Infrastructure
Rate Pressures
Security
Workforce Issues (Recruitment, Attrition, Retention and Succession Planning)
Water Resources (Finding or Sustaining High-Quality Water Resources)
Expansion of Customer Base (Meeting the Needs of a Growing Customer Base)
Regionalization
Workforce Motivation (Continued Engagement and Appropriate Focus)
Political and Regulatory Environment

PEOPLE ARE THE MOST IMPORTANT

As was learned during the project, the successful application of KM embodies several critical elements: people, culture, process, and technology. While each of these elements contributes to a KM initiative's success, it is the people element, considered in context with the organization's culture, that is most important to achieving positive and sustainable results. Organizations that support and nurture cross-functional communication, teamwork, innovation, and solution-oriented collaboration provide fertile ground for realizing the benefits of KM applications. For example, the loss of experienced personnel due to the aging workforce results in inexperienced replacements taking on increasing responsibility. Utilizing KM strategies to capture, code, and effectively transfer know-how and lessons learned can minimize the competency gap caused by this mismatch of experience to responsibility. Key to developing this effective knowledge transfer are collaborative knowledge harvesting techniques within and across teams and functional groups, complemented by efficient processes and procedures documenting and supporting standard methods for capturing, managing, presenting and accessing organizational knowledge. Unlike private industry, the relatively noncompetitive nature of the drinking water industry indicates that the benefits realized from KM strategies can be accelerated and maximized through the establishment of common KM methodologies and standards.

FOSTER A LEARNING CULTURE NOW

The take-away from this observation is that utilities need not wait for future research results to begin fostering a learning culture and implementing organizational change agents that facilitate establishing a collaborative environment. Importantly, this environment must be designed to promote learning, innovation, horizontal communication, empowerment, and teamwork through efficient processes based on standard and practical approaches. Additionally, Utility Managers should realize that the development of an enabling environment that supports learning is only one aspect that needs attention. To enhance the organization's capabilities relative to adapting to meet the new challenges, the learning that takes place must be processed and converted into "executable know-how" that effectively and efficiently solves problems. Various studies have detailed how organizational learning is not an easy goal to achieve and that organizations have a tendency to forget. A concept that must be considered in the context of KM and one closely aligned with existing research results is "organizational amnesia" (see Appendix C for additional discussion).

CAUTION: TECHNOLOGY IS NOT KNOWLEDGE MANAGEMENT

Unfortunately, technology is sometimes used interchangeably with the term "Knowledge Management." Although some early KM adopters thought implementing a database or other software tool constituted knowledge management, it is now known, through the documentation of case studies and academic research, that the role of technology is one of an "enabler." Results from this project's research indicate that many utilities have implemented and are planning deployments of technologies that are typically used to support KM initiatives. The summary of the survey responses generally portrays a utility environment that has deployed or is planning to deploy KM-enabling technologies such as databases, collaborative software, and document/content management repositories to support the collection, management, and distribution of information. Additionally, many utilities are procuring and deploying large-scale point solutions, such as Asset Management and Customer Information Systems. While it was not surprising to find that information capture and management technologies are being deployed at a high rate by utilities, it was surprising to find that more than half of the participating utilities believed they have between approximately one-third and three-quarters of their critical information types such as drawings, maps, permits, O&M manuals, and standards already in digital format.

From the technology procurement and deployment perspective, the key considerations for utility decision makers are:

- The procurement and deployment of technology will not in and of itself prove effective to solving operational problems, improving performance, or complying with regulations.
- Many times adequate technology already exists in the environment and what is missing is the integrative thought and view which sees the organization as a system.
- This system is composed of the elements of people, culture, processes, and technology that interact with the environment (e.g., customers, regulators, governing bodies, suppliers).
- The outputs of these interactions – if collected, filtered, managed, disseminated, and reused – can enhance a utility's capabilities, knowledgewise.

As can be seen, the key consideration for utility leaders pertaining to this system's viewpoint is much more complex than technology deployment. Such a perspective strives to look at the best structures and methods for aligning the people, culture, and processes while leveraging technology in response to and anticipation of stimuli from the environment (internal and external).

CLOSE THE RESOURCE AND SKILL GAP

Although it is the project team's contention that industry decision makers should pursue incorporating KM strategies into strategic planning processes, the application of KM is not without challenges. When participants were asked to identify the main challenges that their utility would face in implementing a KM pilot within their organizations, the hurdle most often cited dealt with available and appropriately skilled resources. In general, the respondents describe their organizations as spread thin by competing initiatives and struggling with maintaining and developing the proper mix of technical and project management skills. Interestingly, as was seen with the listing of industry challenges discussed earlier, this perspective agrees with feedback from organizations outside the industry. A 2004 survey from *CIO* Magazine of 340 Information Technology (IT) executives regarding IT staffing needs lists demanding workloads and retaining and finding proper skill sets as critical capability shortfalls. Likewise, a recent META Group, Inc. report on project management best practices stated "80% of IT organizations report a lack of enterprise project management capabilities [i.e., large-scale initiatives encompassing multiple business units and or business processes] as being a key IT workforce issue." The fact that skills such as these are not unique to successful KM applications but are critical to the success for any initiative highlights a gap that needs to be addressed. The implication for utilities is that this shortcoming must be addressed even for those organizations not considering implementing KM. The pervasiveness of technology, the high frequency of adding or upgrading business systems, and the increasing importance of these systems to operating a utility present a strong business case for closing this resource and skill gap.

CRITICAL SUCCESS FACTORS

In summary, the drinking water industry can benefit from implementing KM strategies. The drivers and challenges that influenced the adoption of KM as a management strategy in other industries are similar to what the Utility Manager is currently facing. Although implementation hurdles exist, they can be minimized by incorporating the following critical success factors in the KM strategy:

- The core processes of strategic planning, technology master planning, business/operational planning, CIP planning, and HR life cycle activities must be viewed as intertwined and processes must exist to maintain their alignment
- KM initiatives must be not be thought of as independent initiatives but must be viewed as part of the utility's strategy and embedded as part of continuous improvement activities
- KM initiatives must be adequately and properly resourced. Utility Managers looking to implement initiatives supporting KM must investigate and plan for the critical skills and resource requirements for all aspects of the initiatives' life cycle (e.g., planning, implementation, operations).

CHAPTER 5: CANDIDATE RESEARCH PROJECTS

OVERVIEW OF RESEARCH CANDIDATES

The research team, in collaboration with the project participants, identified the following candidates for future research projects. Guiding the development of these candidate projects were the key industry challenges and industry KM definition as identified by the project team. These challenges are discussed in further detail in the following section. These candidates have common objectives of further defining the application of KM for the industry and identifying best practices and practical approaches for the inclusion of KM strategies within the mainstream of utility management "best practices." Ultimately, the goal is to establish KM as an applicable and mainstream management strategy for the industry. Each candidate research project has objectives of identifying, defining and evolving one or more of an industry strategy, standard, or solution that can be leveraged by any utility in support of KM.

1. Strategies to Help Drinking Water Utilities Ensure Effective Retention of Knowledge

Description of Issue

Organizations are experiencing the effects of knowledge attrition. Some of this is the result of the departure of veteran workers from organizations due to early baby boomers reaching retirement age or retiring early. Some of this is due to rapid growth, turnover, acquisitions, and internal deployment. Whatever the reason, the loss of veteran workers may result in a loss of a significant amount of working knowledge, processes, and lessons learned that will affect organizations. In response, organizations are searching for ways to capture, index, and store the tacit and explicit knowledge held by their employees before it is lost. Capturing the knowledge will enable their successors to quickly harvest and retrieve it efficiently, when and where it is needed. Drinking water utilities in particular face the added challenge of attracting and retaining long-term and reliable replacements for their retiring workforce because the next generation of younger individuals is lured away to other industries after a short time, further straining the system and widening the knowledge gap. Thus, the need to establish an effective knowledge retention program within a KM program becomes even more urgent for the drinking water utility.

Benefits for Water Supply Community

Identifying critical success factors and establishing a strategy for an ongoing approach to knowledge retention within a drinking water utility would help ensure that business continuity, overall performance, and customer satisfaction are maintained at high levels, while minimizing the impact of workforce attrition.

Objective

Identify strategies, tools, and techniques that utilities can use to retain knowledge from the retiring and shifting workforce. Identify key drivers, critical success factors, barriers, costs, and benefits related to implementing knowledge retention programs.

Estimated Funding

$300,000

Research Approach

The research approach would support the discovery and identification of strategies, tools, and techniques, used both within and outside of the water supply industry, that have resulted in successful knowledge retention programs. The research approach would result in the identification of key drivers and critical success factors of knowledge retention programs, tools and techniques for the discovery of critical knowledge capture, as well as the identification of a range of costs and benefits for implementation and measures of effectiveness of knowledge retention strategies.

The research approach may include literature review, case study compilations, surveys, interviews, workshops, and other techniques that result in the desired outcome. The final report would document the research approach and would focus on the presentation of the findings so that they are relevant and useful to drinking water Utility Managers desiring to implement knowledge retention programs. The final report would identify and describe a range of strategies, tools, and techniques used to implement knowledge retention programs, and would include a range of cost, benefits, key drivers, critical success factors, and barriers to implementation.

2. Define Industry Taxonomy Standards

Description of Issue

Knowledge Management (KM) is a business strategy and process by which an organization identifies, captures, indexes, manages, and stores human experiences, data and information and provides methods to access the information to support effective decision making, assure compliance, and improve performance, innovation, and business continuity, all on a timely and sustainable basis. Since information (e.g., data, documents, drawings, permits, contracts, requests for proposals [RFPs]) will be a core component in any KM initiative, thought needs to be given to the structure and categorization that content owners, contributors, and users will rely on when updating and accessing information and knowledge repositories. Current research suggests that typical knowledge workers spend 4-6 hours per week searching for information, and that 50% of these searches fail to locate the desired information. The results of this reality are that work is duplicated, inaccurate versions of documents/drawings are used to make decisions, and project timelines are negatively impacted.

Benefits for Water Supply Community

Development of an industry-specific taxonomy that provides a standard tool kit for jump-starting a utility's KM project would reduce the time that the project team would spend on designing information classifications, relationships, and search-and-retrieval properties. When applied, the standard would provide the basis for cataloging, storing, retrieving, and presenting knowledge within the utility. The standard would reduce utility KM implementation complexity and costs while compressing required implementation time frames. Additionally, the industry standard would recognize external stakeholders such as suppliers, boards, and customers and provide structures for supporting their unique requirements.

Objective

Identify and develop an industry taxonomy and information architecture standard which can be leveraged as a KM-enabler for utilities.

Estimated Funding

$175,000 - $250,000

Research Approach

The research approach would require active involvement of the drinking water industry and may include literature reviews, case studies, interviews, benchmarking efforts, surveys, workshops, pilot studies, or any other approach that would support the identification of best practices associated with designing and implementing information architectures utilizing taxonomies. The research approach would result in:

- Identifying organizations both within and outside of the drinking water industry that have developed information taxonomies in support of KM strategies
- Identifying the tasks and processes necessary for developing taxonomies
- Identifying critical success factors necessary for successful taxonomy development
- Identifying common business vocabularies, glossaries, and indexes in support of the industry's content
- Investigating organizational structure, roles and responsibilities that support successful taxonomy projects
- Development of a tool kit containing a base taxonomy and taxonomy development road map.

The final outcome would be a resource for drinking water utilities to use to develop taxonomies in support of their knowledge management strategies. The final report would present the findings in a format that is relevant and useful to drinking water Utility Managers that are considering developing information architectures in support of KM initiatives.

3. Utility Organizational Development to Facilitate Knowledge Management Business Strategies

Description of Issue

The early adopters of KM experienced many barriers to the successful implementation and operation of initiatives supporting their KM strategies. In many cases the root cause for these barriers was based in the failure by leaders and KM implementation teams to consider the organizational development aspects of implementing a strategy such as KM. In deciding to move forward with KM, the Utility Manager is really saying that a business transformation is underway. Without investigating and considering the change management and behavioral dimensions with respect to collaboration and knowledge sharing, risks pertaining to the successful implementation of KM are either understated or ignored. Additionally, potential barriers to success with respect to people and cultural change management are not proactively considered.

Benefits for Water Supply Community

Documented best practices, strategies, and techniques would help align organizational development considerations with KM strategy execution. Documented risk factors relative to KM related organizational development considerations and strategies for mitigating those risks would be provided. A focus of the research will be to develop a tool kit providing Utility Managers with strategies and approaches that can be adapted to the unique environments and cultures of their organizations. Specifically, this initiative addresses the organizational development (collaboration, knowledge diffusion, communication, change management, learning environment, trusting climate, incentive strategies [rewards and recognition], and so on) aspects of this business transformation.

Objective

Develop a tool kit for drinking water utilities to evaluate and develop their organizational cultures so that they are supportive of successful implementation of knowledge management initiatives.

Estimated Funding

$250,000

Research Approach

The research approach would require active involvement of the drinking water industry and may include literature reviews, case studies, interviews, benchmarking efforts, surveys, workshops, pilot studies, or any other approach that would support the identification of sound techniques and best practices associated with implementing KM and creating and sustaining a knowledge-sharing and learning organization. The research approach would result in:

- Identifying organizations both within and outside of the drinking water industry that have implemented KM and targeted developing learning cultures as strategic and documenting their experiences in creating and maintaining them
- Identifying strategies and approaches that facilitate knowledge-sharing and diffusion with and without enabling technology
- Investigating solutions for facilitating knowledge sharing and transfer that are not technology-intensive. This will support utilities with limited technology budgets, misaligned technology skills, and personnel who do not have ready access to applicable technology tools
- Discovering approaches, techniques and best practices from current organizational development research to leverage the behavioral science knowledge pertaining to achieving broad-scale culture change
- Investigating the role of an organization's leadership in planning for, overcoming resistance to, and successfully implementing organization structures (e.g., new roles, rewards, and recognition norms) supporting KM strategies and initiatives
- Investigating organizational structure, roles and responsibilities that support successful knowledge sharing and learning cultures and barriers to successful implementations of KM in organizations
- Identifying training approaches for successfully transforming organizations into knowledge sharing and learning organizations in support of KM strategies and knowledge sharing.

The final outcome would be a body of knowledge for drinking water utilities to use in order to prepare their organizations for transitioning to a knowledge-enabled organization. The final report would present the findings in a format that is relevant and useful to drinking water utility leaders considering creating and nurturing sustainable learning organizations. It would identify a range of implementation strategies, business practices, critical success factors, and techniques from academic research and practitioners supporting successful change management in support of KM. It would also identify common barriers to success and strategies for overcoming these barriers.

4. Best Practices for Aligning Core Planning Processes Necessary for Successfully Applying Knowledge Management

Description of Issue

It is hypothesized that misalignment of a utility's core planning processes (i.e., strategic planning, operational planning, CIP planning, HR life cycle planning, and technology master planning) would most likely result in KM initiatives with significant cost overruns, being abandoned in mid-stream, or KM solutions that cannot be adequately maintained in production.

Benefits for Water Supply Community

Implementation risks for KM strategies associated with poorly aligned plans would be mitigated. Utility Managers would understand the relationships and dependencies of these critical core planning processes and be in a position to proactively address issues that, if left unattended, would negatively impact the execution of KM strategies.

Objective

Investigate the relationship of a utility's core planning processes to implementation of a successful KM program. Identify how misalignment or an underdeveloped process impacts the successful deployment of a KM strategy and its resulting initiatives.

Estimated Funding

$250,000

Research Approach

The research approach would support the discovery and identification of strategies, tools, and techniques, used both within and outside of the water supply industry, that have resulted in successful alignment of core business planning processes. The research approach would result in the identification of key drivers and critical success factors for successfully integrating the core planning processes that are critical to the successful implementation of KM.

The research approach may include literature review, case study compilations, surveys, interviews, workshops, and other techniques that result in the desired outcome. The final report would document the research approach and would focus on the presentation of the findings such that they are relevant and useful to drinking water Utility Managers desiring to implement knowledge management programs. The final report would identify and describe a range of strategies, tools, and techniques helpful in maintaining alignment of these core planning processes such that they support the utility's KM strategy.

5. Identification of Required Supporting Resources and Infrastructure to Enable the Knowledge-Based Utility

Description of Issue

Lessons learned from documented case studies pertaining to applied KM strategies identify critical elements with respect to the technical infrastructure, skills, and resources required to implement and support KM-driven initiatives. The Utility Manager needs to be aware of all of the required components (e.g., hardware, software, resource, skills) that must be available to adequately support implementing and maintaining KM strategies. Successful KM implementations typically involve enabling technologies that require resources and skills that the Utility Manager may not currently have "in-house" or skills/resources that are already allocated.

Benefits for Water Supply Community

The development of a comprehensive road map and framework that describes all of the critical components of infrastructure and supporting resources for designing, supporting and enhancing KM-based solutions would provide the Utility Manager and KM implementation teams with an accurate perspective of KM's impact on their infrastructure and supporting resource pool. Key benefits and outcomes of this research would be the identification of technologies (e.g., search tools) that could access and bridge existing technology applications and infrastructures as well as a template to be used in developing a sound and realistic business case for initiatives supporting KM strategies. Additionally, as Utility Managers continue to procure infrastructure components in support of enabling core business processes, consideration for leveraging this new infrastructure to support KM strategies would be contemplated.

Objective

Develop an enterprise integration model depicting an integrated flow (i.e., presentation, search, retrieval) of both structured (database) and unstructured (office applications) data and information contained in utility business systems and processes. Develop a supporting template providing utilities with the ability to readily understand the critical components (e.g., organizational skills/roles [technical and business], measurement approaches, readiness assessment, and security/business continuity/operations) that are necessary to support the implementation and maintenance of a KM strategy.

Estimated Funding

$250,000

Research Approach

The research approach would require active involvement of the drinking water industry and may include literature reviews, case studies, interviews, benchmarking efforts, surveys, workshops, pilot studies, or any other approach that would support the identification of all of the critical components (infrastructure, resource, skills, training) necessary for successful KM strategy implementation. The research approach would result in:

- Development of a KM integration model for typical utility business systems and core processes
- Development of a KM template that serves as a road map for implementing and sustaining KM solutions.

The final outcome would be tools for drinking water utilities to use in order to prepare their organizations for transitioning to a knowledge-enabled organization. The final report would present the findings in a format that is relevant and useful to drinking water utility leaders considering implementing strategies for managing the utility's knowledge.

6. Application of Communities of Practice to the Drinking Water Industry

Description of Issue

The industry has a current need to begin a structured process for technology transfer and collaboration. Additionally, utilities are searching for ways to actively learn from others that have expertise in a given domain or in solving problems similar to theirs. Importantly, this process must be accomplished such that geographically dispersed utilities can participate in the dialogue without traveling to specific locations and members of a utility can participate as their schedule allows. One approach that has gained momentum outside of the industry to facilitate technology and domain expertise transfer is the development of Communities of Practice (CoPs). Research and case studies from leading organizations such as APQC and the National Aeronautics and Space Administration (NASA) has shown that CoPs are important to the efforts of creating, filtering, and sharing knowledge within and across functional groups. For this reason, CoPs are becoming increasingly important to Knowledge Management (KM) initiatives and are viewed as a critical success factor in implementing and sustaining processes needed to move organizations toward a learning culture.

Benefits for Water Supply Community

Development and implementation of CoPs to support collaboration and knowledge transfer is a practical and tactical method for the industry to begin benefiting from KM strategies and concepts. It can provide quick impact to the process of knowledge sharing and diffusion. Establishing effective CoPs would facilitate technology transfer and collaboration among subject matter experts within the industry, and would enhance the flow and collection of information and knowledge across utilities, providing the industry with a collective body of knowledge that can be accessed and leveraged by anyone, anywhere to solve current business/operational problems. The utilization of CoPs by utility personnel will assist the industry in reducing costs, increasing efficiencies, and facilitating the use of best practices across the industry.

Objective

Investigate the application of Communities of Practice to the drinking water community. Design, develop, and pilot a CoP for the drinking water industry and document findings relative to identified success criteria. Identify costs and benefits, key drivers, critical success factors, and barriers to successful implementation and industry adoption of CoPs.

Estimated Funding

$300,000

Research Approach

The research approach would require active involvement of the drinking water industry and would support the investigation of the application of CoPs to the drinking water industry. The research approach would result in:

- Identifying organizations both within and outside of the drinking water industry that have an established track record of successful CoP implementations
- An understanding of a CoP's role within a knowledge-based organization
- Identifying critical success factors, implementation strategies and best practices for implementing CoPs
- Investigating the role of an organization's or industry's leadership and culture in implementing and supporting CoPs
- A set of metrics capable of measuring and providing feedback relative to a CoP's contribution to the organization and industry
- Identifying tools, techniques, and mechanisms for facilitating communications among CoP members and other key stakeholders
- Identification of roles, processes (e.g., peer review of content), skills, and responsibilities necessary to design, develop, and maintain a CoP.

The final outcome would be the implementation of a CoP pilot project. Additionally, a guide will be developed for drinking water utilities to use in understanding the value of CoPs to the industry and how to initiate and nurture CoPs to obtain organizational benefits and complement KM strategies. The final report would present the findings in a format that is relevant and useful to drinking water Utility Managers that are considering developing their own KM business cases or implementing their own KM programs. The final report would identify a range of implementation strategies, best practices, critical success factors, value propositions, and performance metrics related to using CoPs to foster the broadest possible participation and industry involvement in support of technology transfer and CoP efforts.

7. Pilot Business-Enabling Core Utility Information Solutions

Description of Issue

Much of the knowledge that a utility is trying to capture, manage, and reuse is embedded in a small number of key business processes and information life cycles. Although these core processes and the problems they present have common traits across all utilities regardless of geographic location and size, each utility addresses solutions in support of these core processes in unique ways. While in many cases the unique solutions serve the utility well, the lack of standardization and common framework does not allow the industry as a whole to leverage best practices and lessons learned. As a result of this current approach, the industry and individual utilities are negatively impacted by:

- Increased efforts and costs for capturing, managing and distributing knowledge
- Varying degrees of success related to managing knowledge
- Due to lack of standard solutions there exist lost opportunities for cost savings and efficiencies for solution planning and implementations
- Discrete stove-pipe solutions for the same problem across the industry, adding to an increasing mosaic of technology, resulting in increased O&M costs.

Benefits for Water Supply Community

Understanding of how common business problems and core processes can be addressed in a standardized manner would result in efficient, effective, and transportable Knowledge Management (KM) techniques across utilities. This would also provide an industry-wide opportunity for reduced costs when procuring and implementing services and products in support of these common business problems. A key outcome would be a body of knowledge that documents best practices, effective techniques and solutions, and lessons learned for common utility core business processes.

Objective

Identify common industry and core utility processes (e.g., asset management, drawing management, O&M manuals, etc.) that can benefit from KM-based solution approaches. Investigate and document existing "best-in-class" approaches in the industry for these common processes and develop a comprehensive "best practice" model for common core information life cycle management approaches that can be leveraged across the industry by individual utilities.

Estimated Funding

$250,000

Research Approach

The research approach would require active involvement of the drinking water industry and may include literature reviews, case studies, interviews, benchmarking efforts, surveys, workshops, pilot studies, or any other approach that would support the identification of best practices associated with designing and implementing solution architectures supporting core utility processes. The research approach would result in:

- Defined set of core processes that can benefit from KM-based solution approaches
- Documented "best-in-class" approaches for core processes
- Documented critical success factors, implementation approaches, and evaluation criteria
- Demonstrable pilot(s) exhibiting the best practice for embedding KM concepts in one or more of the core processes (e.g., Asset Management, Drawing Management, etc.).

The final outcome would demonstrate to Utility Managers how best to implement core business process solutions that are not confined to an individual project or initiative, but rather are derived from an embedded strategic approach supporting KM strategies with respect to knowledge capture, access, management dissemination, and reuse. The final report would present the findings in a format that is relevant and useful to drinking water utility leaders considering strategies for managing the utility's knowledge.

CHAPTER 6: READINESS ASSESSMENT

OVERVIEW

The findings from this project strongly suggest that the challenges currently facing utilities are similar to influences that motivated organizations outside the industry to investigate, develop, and deploy initiatives in support of KM strategies. The similarity of these challenges indicate that all water utilities are subject to internal and external forces that can be alleviated, controlled, better understood, anticipated, and overcome by leveraging the benefits derived from a knowledge-enabled workforce.

The continuing pressures of regulatory compliance, workforce demographics, aging infrastructures, budget constraints, and increasing stakeholder expectations will motivate utility leaders to investigate further the use of KM strategies as a means for improving their organization's ability to learn, innovate, increase effectiveness, increase efficiencies, and improve their decision-making capabilities. But how does a leader know when to consider implementing a strategic change initiative such as KM, which embodies the critical organizational structures of, people, culture, processes, and technology? To provide a practical means for gauging the utility's readiness for changes in the way that personnel interact, learn, communicate, and how information flows across related business processes, a methodology and model must be developed. Importantly, to be of the most use, the model should be comprehensive in that all of the stages within the KM life cycle must be supported.

KM MATURITY MODEL

The research team has developed a draft model by which the industry and AwwaRF can further evolve a KM strategy in a formalized, deliberate, and measurable manner. This model considers all of the major phases that are embodied in the complete life cycle of KM and provides a road map for growing the organization into a learning culture. Using an approach such as this, a utility can measure its progress with respect to KM, and predict and plan for the organizational changes that must occur to progress along the KM capability and maturity spectrum.

The KM Maturity Model is described in Figure 6.1 with each phase detailed in the subsequent sections. It should be noted that the prospective journey from Vision and Readiness Assessment through Mature and Evolved is depicted in a rather circuitous, nonlinear manner. This is meant to imply that the underlying and unique characteristics of each utility, imposed by all of its internal and external constraints, opportunities, and dynamics, manifest in an evolution which is not always that straightforward, and is generally replete with many adjustments and modifications enacted along the way. Hence, the particular eventual path is as individual as the utility that undertakes the migration.

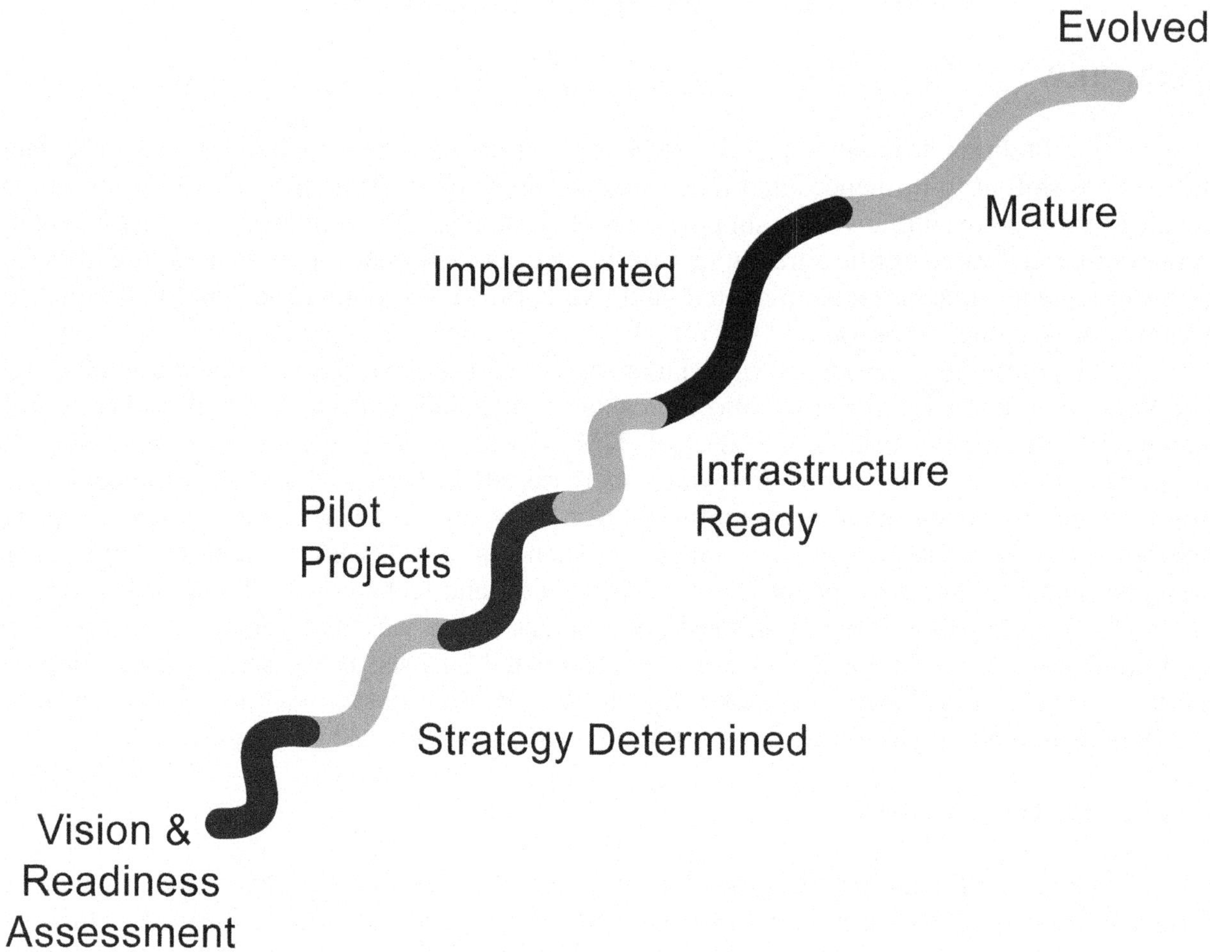

Figure 6.1 Knowledge management maturity model

Vision and Readiness Assessment

This is the first and also most important stage as it establishes the vision for what the organization wants to be with respect to KM and allows utilities to understand their strengths, weaknesses, and opportunities along multiple dimensions that are critical to successfully implementing a KM strategy and supporting initiatives. Because this vision-setting establishes the strategic intent, this phase requires commitment from the executive team to sponsor and champion the complete process.

The work performed during the readiness assessment portion of this phase answers critical questions pertaining to the core dimensions that KM embodies: people, process, culture and technology. This phase also initiates a discussion of what constitutes knowledge for the utility and begins a Knowledge Audit process. This assessment phase is described in greater detail later in this report along with an approach to quantifying a utility's readiness for formally pursuing a KM strategy.

Strategy Determined

This stage involves the preparation and planning necessary to begin initiatives supporting KM strategies. The core project team is assembled, and the scope, objectives, and critical success factors are formulated for the initiative. Critical outcomes of this stage are the internalization of what KM means to a utility, consensus on what constitutes knowledge, the process identified for governance of the overall KM program, and the acquisition of strong executive sponsorship for the strategy. KM projects depend upon strong executive leadership to succeed, with this sponsorship being critical to realizing the full benefits of KM and accomplishing the necessary organizational change that must occur to be successful.

Pilot Projects

During this stage of development, targeted pilot projects are executed to solve key business issues and fine-tune the approach for deploying KM. Key activities of this stage include selection of tools, establishment of baseline infrastructure for the pilot projects, and generation of a shared focus for the implementation team. In addition to solving key business issues, the use of pilot projects generates initial return on investment, further proves the business case for KM, increases understanding of applicability, and builds department-wide consensus and buy-in for the required organizational change.

Infrastructure Ready

During this stage, the KM infrastructure is completed in preparation for organizational deployment. The KM environment has been refined based upon the experience of the pilots and the training program is enhanced and expanded to accommodate participants from outside the pilot team. Associated internal support resources are allocated, trained, and functionally ready to support large-scale deployment of KM. Communications that underscore the importance, need, and benefit associated with KM have been drafted and delivered to the organization.

Implemented

An initiative supporting KM has been deployed and is used by 25% of the organization. Use and effectiveness of the system are formally measured and monitored. Training efforts are active and ongoing. KM processes have been determined, implemented and documented. The KM strategy is now a major focus of the utility.

Mature

Collaboration has become the "norm" of the organization. KM is viewed as a key indicator for the organization. Innovation and learning resulting in improved decision-making capability is occurring as a direct result of the initiative. KM is growing dynamically and fueling improvements in key performance indicators of the utility. Additionally, the success of KM is not dependent on any one individual or position but has become embedded in core processes of the utility.

Evolved

KM is fully deployed and integrated. The process has become sustainable and is viewed as a core competency of the utility. KM has become a way to achieve continuous improvement and enhance organizational effectiveness and capability. Utility personnel efficiently receive the right information at the right time in the proper form regardless of role or level in the organization.

READINESS ASSESSMENT DIMENSIONS

After the visioning session and prior to any KM strategy being initiated, the organization should perform an internal assessment with respect to its readiness to implement, support, and sustain initiatives in support of KM strategies. As discussed earlier, KM differs from traditional management approaches in that it strives to create a learning environment in which personnel have on-demand access to an organization's data and information in such a manner that informed decision making and efficient and effective operations are a result. Key characteristics of this environment are that it is knowledge-enabled and consciously incorporates people, processes, and technologies in the development and execution of strategic business plans. Since a successful KM strategy must rely on core organizational structures to be aligned, it is necessary to analyze and evaluate certain dimensions in the context of KM perspectives. The goal of this analysis is to identify areas that have deficiencies with respect to critical success factors and criteria necessary for a successful KM strategy. Addressing these deficiencies prior to embarking on a KM strategy will reduce the risk of failure and lay a strong foundation for evolving into a sustainable knowledge-enabled organization. These key structural dimensions are depicted in Figure 6.2.

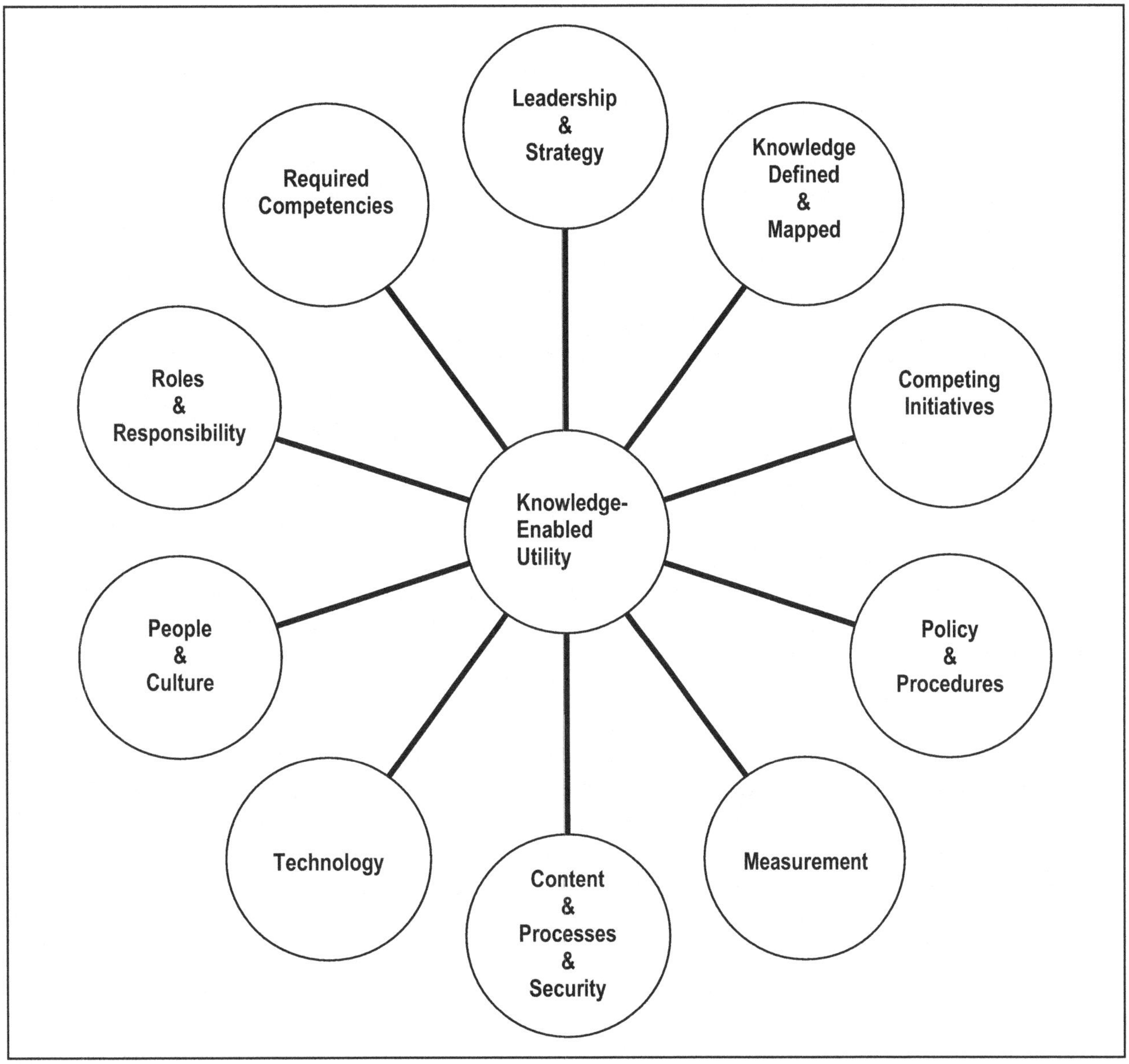

Figure 6.2 KM readiness assessment dimensions

A FINAL WORD OF ENCOURAGEMENT

A journey of a thousand miles begins with a single step. (Lao-tzu)

The foregoing readiness assessment process is intended as a useful tool whereby any given utility can evaluate how well it is equipped to launch, carry out, and sustain a KM strategy and related supporting initiatives.

But even without employing the model, the implementation of a KM initiative is often less daunting than one would at first surmise, as evidenced by many of the experiences related by project participants. Several had already (and perhaps inadvertently) embarked on one or more initiatives that could be classified as "Knowledge Management." These utilities did not formally name the initiative "Knowledge Management"; did not have dedicated KM project line items in their budgets; and did not have a ranking KM champion on board to tout the inherent benefits and solicit stakeholder buy-in.

And yet, compiling As-Built drawings and O&M manuals, for example, and making such documents readily accessible to interested and appropriate parties, certainly qualify as valid KM endeavors, at least on a rudimentary level, and make undeniably compelling arguments for providing additional examples. Importantly, few would dispute the numerous advantages of such pursuits. Hence, much of the early significant groundwork in promoting and recognizing the value of implementing KM initiatives is already under way, making subsequent endeavors that much easier to manage and realize, and hopefully instilling within the organization an eagerness for more.

The holistic context of the framework assures that all of the core elements of KM – people, culture, process, and technology – are consciously considered and evaluated throughout the KM adoption life cycle. For organizations that have already started the process of realizing a knowledge-enabled utility, the addition of the Readiness Assessment Framework and KM Maturity Model described above provides a mechanism for continuing to scale their efforts in a structured manner with a focus of moving toward a learning organization model. For utilities that are just beginning, the framework and model facilitates overcoming the inertia and provides a road map that can be used to guide the utility leadership along the journey.

APPENDIX A: SURVEYS

The project team conducted three surveys of its participating utilities over the course of several months. Survey #1 obtained baseline information related to the demographics of project participants and characteristics of their respective utilities. Survey #2 probed metrics used to manage the utility; cultural characteristics and the utility's focus on Capital Improvement Program (CIP) initiatives; and the utility's approach and process for managing CIP initiatives. Survey #3 identified what knowledge-enabling technologies currently exist or are planned to be implemented by the participant's utilities, and collected information on potential hurdles to implementing a knowledge management pilot.

The first survey was conducted by telephone to help establish a personal relationship among team members. The second and third surveys were conducted on-line through the project's Web site for the added convenience of participants, and were occasionally followed up by telephone inquiry for added clarification and detail.

SURVEY #1

In support of AwwaRF's research project, "A Knowledge Management Approach to Drinking Water Utility Business," the following questionnaire surveys information about you and your organization, your initial KM experiences and impressions (if any), and your desired outcomes from your participation in this project. Your cooperation in completing this survey is appreciated.

Respondent Demographics

1. Please check the box that most closely corresponds to your job level:

☐ Executive Director
☐ Assistant Director
☐ Manager
☐ Line Supervisor
☐ Other: ______________________________

2. Please check the box that most closely corresponds to the functional area you belong to:

☐ Engineering
☐ Executive Team
☐ Facilities Management
☐ Finance
☐ Human Resources
☐ Information Systems
☐ Maintenance
☐ Operations
☐ Purchasing
☐ System Planning
☐ Other: ______________________________

3. Check the box that corresponds to the number of years in your current position:

☐ <1 ☐ 1-3 ☐ 4-10 ☐ 11-20 ☐ >20

4. Check the box that corresponds to the number of years in your current organization:

☐ <1 ☐ 1-3 ☐ 4-10 ☐ 11-20 ☐ >20

5. Please check all of the annual conferences listed below that you typically attend:

□ AWWA ACE/IMTech
□ AWWA (Section)
□ AMWA
□ Industrial Water Conference
□ WEF/WEFTEC
□ WEF (State)
□ AMSA
□ Other: ______________________________

Utility Characteristics

6. Please check the box that best describes the type of Utility that you represent:

□ Public □ Private

7. Please check all applicable boxes below that describe the services your Utility offers:

□ Water Retail □ Water Wholesale □ Wastewater Treatment □ Wastewater Collection
□ Electrical □ Other:_____________________________

8. Please check the box that corresponds to the number of employees in your Utility:

□ <100 □ 100-249 □ 250-499 □ 500-749 □ 750-1000 □ >1000

9. How many residential water customers (i.e. total population) does your Utility service?

□ <75K □ 75K-249K □ 250K-499K □ 500K-999K □ 1M-1.99M □ >2M
□ N/A

10. How many residential wastewater customers (i.e. total population) does your Utility service?

□ <75K □ 75K-249K □ 250K-499K □ 500K-999K □ 1M-1.99M □ >2M
□ N/A

11. What is the average water treatment capacity of your Utility?

□ <10 MGD □ 10-24 MGD □ 25-49 MGD □ 50-99 MGD □ 100-249 MGD
□ 250-499 MGD □ 500-749 MGD □ 750-999 MGD □ >1 BGD □ NA

12. What is the peak water treatment capacity of your Utility?

□ <10 MGD □ 10-24 MGD □ 25-49 MGD □ 50-99 MGD □ 100-249 MGD
□ 250-499 MGD □ 500-749 MGD □ 750-999 MGD □ >1 BGD □ NA

13. What is the average wastewater treatment capacity of your Utility?

□ <10 MGD □ 10-24 MGD □ 25-49 MGD □ 50-99 MGD □ 100-249 MGD
□ 250-499 MGD □ 500-749 MGD □ 750-999 MGD □ >1 BGD □ NA

14. What is the peak wastewater treatment capacity of your Utility?

□ <10 MGD □ 10-24 MGD □ 25-49 MGD □ 50-99 MGD □ 100-249 MGD
□ 250-499 MGD □ 500-749 MGD □ 750-999 MGD □ >1 BGD □ NA

15. What is the size of your service area in square miles?

16. What value below best represents your total utility annual operations (non-capital) budget?

□ <$10M □ $10M-50M □ $50M-100M □ $100M-150M □ $150M-200M □ >$200M

Key Organizational Issues and Pressures

17. In your opinion what are the key issues that your utility is facing. Identify and check the three most important factors from the list below.
(1 = most important, 2 = the second most important, 3 = the third most important)

_____ Increased Regulations

_____ Reorganization

_____ Workforce Attrition

_____ Workforce Recruitment/Succession Planning

_____ Workforce Retention

_____ Increased Competition

_____ Mergers and Acquisitions

_____ Privatization

_____ Regionalization

_____ Aging Infrastructure

_____ Security of Systems and Infrastructure

_____ Rate Pressures

_____ Collections

_____ Expansion of Customer Base

_____ Contraction of Customer Base

_____ Energy Use and Management

_____ Water Resources

_____ Other:____

Tools Currently Used

18. From the list below check all of the tools that you or your utility currently uses in performing daily job functions.

You	Your Utility	
□	□	E-mail
□	□	Voice Mail
□	□	E-Learning Software
□	□	Web-Based Discussion Forums
□	□	E-Meetings (e.g. WebEx, Placeware)
□	□	Teleconferencing/Video Conferencing
□	□	Report Generation Tools (e.g. Crystal Reports)
□	□	Hard Copy or Electronic Reports
□	□	Spreadsheets
□	□	CAD Software
□	□	CAD View and Markup Software
□	□	Internal Web sites / Intranet / Internet / Portals
□	□	Documents / Images / Records / Drawing Management
□	□	Maintenance Management / Enterprise Asset Management
□	□	Customer Billing / Customer Information / Customer Relationship
□	□	Project Management Software
□	□	Geographic Information System (GIS)
□	□	Laboratory Information Management System (LIMS)
□	□	SCADA / DCS
□	□	Finance / HR / Procurement (ERP)
□	□	Laptop / Desktop Computer
□	□	PDA
□	□	Other: ______________________________
□	□	N/A

Participant's Personal Experiences and Expectations

19. In the space below, describe what you believe Knowledge Management means. If at this time you do not have an opinion of what Knowledge Management means, please comment accordingly.

Knowledge Management is…

20. If you have had previous experience with a Knowledge Management project or initiative, from the list below please choose the appropriate role and or forum for your experience. If you have had multiple experiences, please choose all that apply.

- □ N/A
- □ Project Manager of an implementation
- □ Exposed to the term at a conference/seminar/class
- □ Have read articles discussing Knowledge Management topics
- □ Member of a Project Team implementing a Knowledge Management project/initiative
- □ Owner of a Business Process supported by a Knowledge Management project/initiative
- □ Other: Please describe below.

My previous experience with a Knowledge Management project/initiative was…

21. What motivates you to actively participate in an initiative covering the Knowledge Management (KM) topic? Check all that apply.

- □ I understand what KM is and believe a KM strategy can benefit our utility.
- □ I understand what KM is and would like to learn how KM can benefit our utility.
- □ I want to understand how other organizations have used KM to solve problems.
- □ I know little about KM and want to learn more about it.
- □ I know my organization does not effectively manage information and I think KM may act as an enabler to improve collaboration within teams and reuse of information.

22. In the space below, describe your expectations from a personal perspective of what you hope to gain by participating in this initiative.

My expectations from a personal perspective are…

23. In the space below, describe your expectations from an organizational perspective of what you hope your organization gains from participating in this initiative.

My expectations from an organizational perspective are…

SURVEY #2

The purpose of this survey is to gain additional information regarding today's utilities and is divided into three sections. The survey should take 15-20 minutes to complete and participants are requested to complete the survey on-line. The research team will contact participants individually after completion to discuss answers and gain additional insight from participants.

The first section inquires about how business metrics are used to guide utility operations and assess performance. The second and third sections capture information concerning organizational cultures and planned initiatives.

Section One: Business Metrics

1. **What are some examples of key measures/performance indicators your utility relies upon to measure the effectiveness of your business unit or of the utility?**

2. **Are you personally familiar with the term "benchmarking"?**
   - ☐ Yes
   - ☐ No

3. **Does your utility prepare, participate in, or perform any "benchmarking" activities?**
   - ☐ Yes
   - ☐ No

   If yes, please specify:

4. **If you use benchmark data, is the process**
   - ☐ Formal
   - ☐ Informal

     How frequently is the data collected?
     How is it interpreted and reviewed?

5. **Are you aware of the QualServe benchmarking initiative led by AWWA?**
   - ☐ Yes
   - ☐ No

Section Two: Organization Culture

6. **Are employees encouraged to share with others their knowledge and expertise?**
   - ☐ Yes
   - ☐ No

7. **Do any formal or informal mechanisms promoting collaboration among peers/teams/divisions/sections/groups exist within your organization?**
   - ☐ Yes
   - ☐ No

   If yes, please specify:

8. **Is failure viewed as an opportunity to learn within your organization?**
   - ☐ Yes
   - ☐ No

9. **Are there any policies, procedures, or processes in place to capture and address lessons learned, best practices, and subject matter expertise?**
   - ☐ Yes
   - ☐ No

   If yes, please explain.

- Are utility documents (drawings, specifications, as-builts, photographs, correspondence, etc.) currently managed effectively allowing retrieval and referral?

- Centralized or decentralized within the utility?

- Electronically or hard copy (physical)?

Section Three: Planned Initiatives

10. What is your Capital Budget for the upcoming or current year?

11. With respect to pending/planned initiatives, what do you consider to be your utility's primary area(s) of focus.

12. "Generally our utility completes projects consistent with the identified scope, on time, and within budget?"

- ☐ Strongly agree
- ☐ Agree
- ☐ Neither agree or disagree
- ☐ Disagree
- ☐ Strongly Disagree

Would your answer be different if the question differentiated between:

- Facilities Construction
- Technology Implementation

13. "Our organization follows established standards and methodologies, and applies a disciplined approach to executing initiatives?"

- ☐ Strongly agree
- ☐ Agree
- ☐ Neither agree or disagree
- ☐ Disagree
- ☐ Strongly Disagree

14. "Our organization has an effective process in place to nominate, review and approve proposed initiatives."

- ☐ Strongly agree
- ☐ Agree
- ☐ Neither agree or disagree
- ☐ Disagree
- ☐ Strongly Disagree

SURVEY #3

This survey is intended to identify currently deployed and planned technologies that are typically needed to support and enable knowledge management initiatives. The primary goal of the survey is to establish the organization's current capabilities with respect to supporting a knowledge management pilot initiative. This survey should be completed by a member of the organization that has intimate knowledge of the current technology environment and ongoing and planned information technology initiatives/procurements.

Database Software

1. **From the list below, please check the Database products that are currently deployed. Please check all that apply.**
   - ☐ Oracle
   - ☐ MS SQL Server
   - ☐ DB2
   - ☐ Sybase
   - ☐ Other. Please specify:

2. **From the list below, please check the Database products that are planned to be deployed in 2005. Please check all that apply.**
   - ☐ Oracle
   - ☐ MS SQL Server
   - ☐ DB2
   - ☐ Sybase
   - ☐ Other. Please specify:

Report Generation/Business Intelligence Software

3. **From the list below, please check the Report Generation and Business Intelligence (e.g., data mining, analytics) products that are currently deployed. Please check all that apply.**
   - ☐ MicroStrategy
   - ☐ Information Builders
   - ☐ SAS
   - ☐ Oracle
   - ☐ Business Objects
   - ☐ Other. Please specify:

4. **From the list below, please check the Document/Content Management and Imaging products that are currently deployed. Please check all that apply.**
   - ☐ MicroStrategy
   - ☐ Information Builders
   - ☐ SAS
   - ☐ Oracle
   - ☐ Business Objects
   - ☐ Other. Please specify:

<u>Document/Content Management/Drawing Management and Imaging Software</u>

5. **From the list below, please check the Document/Content Management and Imaging products that are currently deployed. Please check all that apply.**
   - ☐ AutoDesk Buzzsaw
   - ☐ Bentley ProjectWise
   - ☐ FileNet
   - ☐ Documentum
   - ☐ Hummingbird
   - ☐ Indus Curator
   - ☐ OnBase
   - ☐ Stellent
   - ☐ Kofax
   - ☐ Captiva
   - ☐ Other. Please specify:

6. **From the list below, please check the Document/Content Management and Imaging products that are planned to be deployed in 2005. Please check all that apply.**
   - ☐ AutoDesk Buzzsaw
   - ☐ Bentley ProjectWise
   - ☐ FileNet
   - ☐ Documentum
   - ☐ Hummingbird
   - ☐ Indus Curator
   - ☐ OnBase
   - ☐ Stellent
   - ☐ Kofax
   - ☐ Captiva
   - ☐ Other. Please specify:

Collaborative Software

7. **From the list below, please check the Collaborative Software products that are currently deployed. Please check all that apply.**
   - ☐ NetMeeting
   - ☐ WebEx
   - ☐ Placeware
   - ☐ eRoom
   - ☐ Lotus Notes
   - ☐ Discussion Databases
   - ☐ Video Conferencing
   - ☐ Other. Please specify:

8. **From the list below, please check the Collaborative Software products that are planned to be deployed in 2005. Please check all that apply.**
   - ☐ NetMeeting
   - ☐ WebEx
   - ☐ Placeware
   - ☐ eRoom
   - ☐ Lotus Notes
   - ☐ Discussion Databases
   - ☐ Video Conferencing
   - ☐ Other. Please specify:

Digitized Information Assets

9. **From the list below, please check the information assets that are currently in electronic format. Please check all that apply.**
   - ☐ Engineering Drawings
   - ☐ Engineering Standards
   - ☐ O&M Manuals
   - ☐ Water Line Maps
   - ☐ Sewer Line Maps
   - ☐ Permits
   - ☐ Contracts
   - ☐ Other. Please specify:

10. From the list below, please check the information assets that are being planned for conversion from hard copy to electronic format in 2005. Please check all that apply.

- ☐ Engineering Drawings
- ☐ Engineering Standards
- ☐ O&M Manuals
- ☐ Water Line Maps
- ☐ Sewer Line Maps
- ☐ Permits
- ☐ Contracts
- ☐ Other. Please specify:

Percentage of Digitized Information Assets

11. From your perspective, what percent range below best describes your utility's current environment as it relates to having information in electronic format vs. paper?

- ☐ 0% - 25%
- ☐ 26% - 33%
- ☐ 34% - 50%
- ☐ 51% - 75%
- ☐ >75%

Web Environment

12. Please describe the Web environment at your utility. Explain the technology (i.e., Servers, OS, HTTP Server, LDAP, etc.) that supports it, the information content that the Web environment delivers to the business, and the business functions that are the primary users. Also describe any use of directory services to support application access management and security objectives. If possible, also comment on how you would describe the level of use of the Web to deliver and distribute critical business information and applications within the utility as a whole.

IS/IT Master Plan

13. Is there currently an IS/IT Master Plan?
- ☐ Yes
- ☐ No

14. If not, are there any plans to create one?
- ☐ Yes
- ☐ No

15. If there is an IS/IT Master Plan, please describe below the process for keeping it aligned with the utility's business plan and objectives.

Knowledge Management Pilot

16. If your utility wanted to implement a pilot that targeted the management of information (e.g., metrics, documents, drawings, reports, etc.) for purposes of increasing the business unit's ability to manage knowledge around its core competencies, from an IS/IT perspective what would be the challenges and hurdles that need to be overcome? Please consider topics such as infrastructure items (e.g., servers, network capacity, etc.), software licenses, technical and project management resources, conversion from hard copy, etc. when answering this question.

APPENDIX B: KNOWLEDGE TRANSFER TECHNIQUES AND THE ROLE OF INCENTIVES

Successfully modifying behavior to support organizational goals of knowledge transfer and knowledge sharing has been a key ingredient for most successful knowledge management programs. There is a large volume of literature dealing with barriers to learning and knowledge transfer and the role that incentives, both intrinsic and extrinsic, play in the success of launching and sustaining knowledge sharing initiatives.

Motivating employees through the use of rewards and recognition is very complex and it is not the intent of this discussion to provide a primer or critique of this topic. Suffice it to say processes to visibly acknowledge knowledge sharing within the organization are a critical success factor to sustaining a KM program. Many experts have observed that while extrinsic rewards (e.g., money) may play a role in the early stages, it is the intrinsic rewards (e.g., peer recognition and acknowledgement) that sustain the KM program.

The information that follows is an introduction to some of the popular techniques that have been and are being used as strategies in support of knowledge transfer and sharing initiatives.

KNOWLEDGE TRANSFER/SHARING TECHNIQUES

After Action Review

A process developed by the U.S. Army to help teams learn quickly from their successes and failures and share their learning with other teams. This process involves conducting a structured and facilitated discussion after a task or project has been completed to review what should have happened, what actually happened, and why it happened; this allows participants to learn how to sustain strengths and improve on weaknesses in subsequent tasks or projects.

Communities of Practice

Networks of people who work on similar processes or in similar disciplines, and who come together to develop and share their knowledge in that field for the benefit of both themselves and their organization(s). They may be created formally or informally, and they can interact online or in person.

Expertise Directory, Experts Directory, Skills Directory, White Pages

A staff directory in the form of a database that includes details of people's skills, knowledge, experience, and expertise so that users can search for people with specific know-how.

Knowledge Harvesting

Knowledge harvesting is a technique that allows the tacit knowledge or know-how of experts and top performers in an organization to be captured and documented. This know-how can then be made available to others in various ways such as through training programs, manuals, best practices, and other forms of knowledge repositories.

Mentoring

Mentoring is a one-to-one learning relationship in which a senior member of an organization is assigned to support the development of a newer or more junior member by sharing his or her knowledge, experience, and wisdom with them.

Peer Assist

A process pioneered by BP-Amoco, in which an individual or team calls a meeting or a workshop in order to tap the knowledge and experience of others before embarking on a project or activity.

Storytelling

The use of stories in organizations as a way of sharing knowledge and helping learning. Stories can be very powerful communication tools, and may be used to describe complicated issues, explain events, communicate lessons learned, or bring about cultural change.

BEHAVIOR-CHANGING INCENTIVES

Monetary Rewards

In much of the literature this type of reward is termed *extrinsic* and is many times based on a metric that an individual might meet with respect to supporting knowledge transfer initiatives. For example, a bonus might be paid out to employees based on the frequency that one had with contributing content to lessons learned repositories or accessing a repository such as this. The literature does discuss scenarios where this type of incentive did not work as planned; in a few identified instances the quality of the submissions was not very good, and people would tend to wait until right before the measurement period ended (e.g., the end of a quarter) to submit content to the repository. Thus the value of having access to the lessons learned or communicating best practices to others in the organization was minimized or negated.

Peer Recognition and Organizational Acknowledgement

In much of the literature this type of reward is termed *intrinsic* and is viewed as critical to sustaining knowledge sharing and embedding it within the culture of an organization. Here again the literature cites various examples of recognizing knowledge-sharing behavior as a way to reinforce this behavior as accepted, expected, and applauded. This recognition ranged from formal and informal communications and acknowledgement for individuals and teams that actively and voluntarily participated in knowledge-sharing activities. As another example, one company in the nuclear industry has created new roles in the organization for personnel identified to have a particular subject matter expertise. These new roles carry a title that communicates that these individuals have a knowledge that makes them recognized as experts in the organization and within their community of peers. This organization has also instituted a mentoring program in which the mentors and individuals being mentored sign a contract stating that the mentor commits to transferring knowledge and the individual being mentored commits to accepting and receiving the knowledge. This is one way that this organization hopes to capture the tacit knowledge held inside these veteran and highly savvy employees.

The preceding information was compiled from the following Web sites and articles:

Andrews, Sue. 2002. Late-Harvested Knowledge [Online]. Available: <http://www.sourceuk.net/indexf.html?02358>. [cited September 12, 2005]

Eisenhart, Mary. 2001. Gathering Knowledge While It's Ripe [Online]. Available: <http://www.kmmag.com/articles/default.asp?ArticleID=209>. [cited September 12, 2005]

Else, Steven. 2003. Practicing Knowledge Communities [Online]. Available: <http://www.kmmag.com/articles/default.asp?ArticleID=1044>. [cited September 12, 2005]

National Electronic Library for Health. 2005. Knowledge Management Glossary [Online]. Available: <http://www.nelh.nhs.uk/knowledge_management/glossary/glossary.asp>. [cited September 12, 2005]

O'Dell, Carla, and C. Jackson Grayson. 2001. Identifying and Transferring Internal Best Practices [Online]. Available: <http://www.kmmag.com/articles/default.asp?ArticleID=271>. [cited September 12, 2005]

Rao, Madanmohan. 2002a. Best Practices in KM [Online]. Available: <http://www.kmmag.com/articles/default.asp?ArticleID=988>. [cited September 12, 2005]

Rao, Madanmohan. 2002b. Emerging Knowledge [Online]. Available: <http://www.kmmag.com/articles/default.asp?ArticleID=1017>. [cited September 12, 2005]

Rao, Madanmohan. 2002c. Siemens Best Practices [Online]. Available: <http://www.kmmag.com/articles/default.asp?ArticleID=1002>. [cited September 12, 2005]

Rao, Madanmohan. 2003. The Deliberate Approach to KM [Online]. Available: <http://www.kmmag.com/articles/default.asp?ArticleID=1078>. [cited September 12, 2005]

APPENDIX C: ORGANIZATIONAL AMNESIA

Arnold Kransdorff's book entitled *Corporate Amnesia: Keeping Know-How in the Company* (Butterworth-Heinemann, 1998) advanced corporate amnesia as the loss of organizational memory which is demonstrated through short and selective memory recall after employees depart. Kransdorff's book cites a study (Warwick University) which concludes that organizations repeat mistakes for two primary reasons. The first stems from the inability to draw from past experiences (time dimension). Somehow these organizations have lost their memory and are incapable of recalling their corporate history. The second reason relates to difficulties that organizations have communicating learned experiences from one part of the organization to another in a timely manner (distance dimension). This study's findings underline two critical dimensions of knowledge management: the need to be able to effectively access collected "know-how" as well as the ability to integrate and disseminate "know-how" across the internal organizational boundaries.

While the key element in overcoming "time-related" organizational amnesia is the creation of an effective form of organizational memory, the approach needed to address "distance-related" (i.e., interdepartmental and/or geographical dispersion) amnesia is different. Measures to overcome "distance-related" amnesia require examination of the processes involved in diffusing knowledge within the organization. An important perspective of "organizational amnesia" is that it establishes that for organizations to be successful in enhancing core competencies, more than information technology hardware and software is required.

Additional information on organizational amnesia may be derived from the following Web sites and articles:

Conklin, Jeff. 2001. Designing Organizational Memory: Preserving Intellectual Assets in a Knowledge Economy [Online]. Available: <http://www.cognexus.org/dom.pdf>. [cited September 12, 2005]

Forman, David C. [no date.] Harnessing the Elusive Asset: Developing Intangible Organizational Capital [Online]. Available: <http://mc1.airscorp.com/portal-upload/superuser/hciGroupPaper_80.pdf>. [cited September 12, 2005]

Kransdorff, Arnold. 1998. *Corporate Amnesia: Keeping Know-How in the Company.* Oxford: Butterworth-Heinemann.

Landry, John R. 1999. Forgetful or Bad Memory? In: *Proc. Thirty-Second Annual Hawaii International Conference on System Sciences,* vol. 1, no. 1, 1069-1076. Piscataway, NJ: Institute of Electrical and Electronics Engineers, Inc. (IEEE).

Othman, R., and N.A. Hashim. 2004. Typologizing Organizational Amnesia. *The Learning Organization: An International Jour.* 11(3):273-284.

Othman, Rozhan, and Noor Azuan Hashim. [no date.] Organizational Amnesia: The Barrier to Organizational Learning [Online]. Available: <http://www.alba.edu.gr/OKLC2002/Proceedings/pdf_files/ID308.pdf>. [cited September 12, 2005]

Politt, Christopher. 2000. Institutional Amnesia: A Paradox of the 'Information Age'? *Prometheus* 18(1):5-16.

APPENDIX D: KEY KNOWLEDGE MANAGEMENT CONCEPTS

Many articles and discussions pertaining to Knowledge Management contain a section dedicated to the definitions of data, information, knowledge, and action, and the relative meaning of these terms to one another. Although differences and perspectives exist on some of the finer points of these topics, there does exist a consistency and consensus regarding several fundamental concepts, including the notion that data, information, and knowledge participate as separate elements within the knowledge model. Additionally, most writers describe a process by which data is transformed into information and information is transformed into knowledge.

Data is a point in time without reference and, although important to the transformation process, it is rarely valuable in itself. Data is transformed into information by applying context, interpretation, or calculation. One way to view this transformation process is to view data as the raw material that is processed into information by being placed into a frame of reference. This process transforms the data into information by placing the data into a business context, resulting in this data being enhanced with meaning and context. In other words, data has no intrinsic meaning by itself and must be sorted, grouped, analyzed, and interpreted before it can be used.

So we can say that information is transformed data that is presented to the user in a meaningful way. This information now has a potential for becoming knowledge if it is known by someone who understands the context and can perceive a pattern because of prior experience, familiarity, developed expertise, intuition, or understanding. Without the proper context, information to one individual is just data to another. Information is transformed into knowledge when it can be used to guide decisions for action. The framework provided by context and insight allow evaluation to be made that leads to confident action and an acceptable response.

As an illustrative example, suppose the numeral 100 (data) is displayed on a Utility Manager's computer screen. Adding the context of "psi" to the numeral begins to transform this raw data into information. Adding a color code and perhaps flashing this information on the screen transforms this information into knowledge for the trained individual who recognizes this as an indication of abnormally high water pressure in a designated low pressure zone with nominal pressures in the 75-90 psi range. Knowing the high pressure alarm has been triggered enables the responsible individual to take an appropriate response action. Here, the raw data has been fully transformed into actionable knowledge by providing the framework of context and insight.

Additionally, two common types of knowledge that most KM discussions contain are *explicit* and *implicit*. Explicit knowledge is the type of knowledge that can be documented and coded. Examples of this type of knowledge are procedures, business rules, project specifications, and as-built drawings. Explicit knowledge is relatively easily communicated and shared.

On the other hand, implicit knowledge, also commonly referred to as tacit knowledge, is typically described as the knowledge contained in an individual's head. This knowledge is highly personal and difficult to document and store in document repositories and knowledge bases due to its intangible qualities and its complexity. Many times the expert who possesses this knowledge is not consciously aware of its value and has an extreme difficulty in articulating it. The impact on the organization is that this knowledge is difficult to access and share, and it leaves the organization at the end of each work day. Cultural and personal issues such as trust and extensive personal contact and peer relationships come into play when trying to capture, organize, and leverage this type of knowledge.

The preceding information was compiled from the following Web sites and articles:

Bierly, Paul E. III, Eric H. Kessler, and Edward W. Christensen. 2000. Organizational Learning, Knowledge and Wisdom. *Jour. Organizational Change* 13(6):595-618.

Bock, Wally. 2004. Postcards from the Digital Age. Knowledge Management that Works [Online]. Available: <http://www.bockinfo.com/041019postcard.htm>. [cited September 12, 2005]

Boisot, Man, and Agusti Canals. 2004. Data, Information and Knowledge: Have We Got It Right? [Online]. Available: <http://www.uoc.edu/in3/dt/20388/index.html>. [cited September 12, 2005]

Hansson, Peter. Knowledge Management (KM) Tools in Mobile Networks – Features, Benefits, Challenges and Characteristics [Online]. Available: <http://www.eknowledgecenter.com/articles/1004/1004.htm>. [cited September 12, 2005]

Luan, Jing, Andreea Serban, Julie Slark, Linda Umbdemstock, and Harriett Robles. 2002. Measuring Learning in the Context of Knowledge Management [Online]. Available: <http://www.cabrillo.edu/services/pro/oir_reports/KM4CCCAll.ppt#1>. [cited September 12, 2005]

Newman, Brian (Bo). 1997. The Siamese Twins Documents and Knowledge. From: Documation '97: Document and Knowledge Management, Santa Clara, CA. West Richland, WA: The Knowledge Management Forum.

Nonaka, Ikujiro. 1994. A Dynamic Theory of Organizational Knowledge Creation. Organization Science, 5(1):14-37.

The OR Society. 2004. Data, Information, Knowledge and Knowledge Management [Online]. Available: <http://www.theorsociety.com/about/topic/projects/notorious/2_2_Data_Info.htm>. [cited September 12, 2005]

Renna, Agostino. The Power of Benchmarking – Transitioning from Data and Information to Knowledge and Action [Online]. Available: <http://www.johnsoncontrols.com/cg-ls/docs/RennaBenchmarkingReprint.pdf>. [cited September 12, 2004]

Stuhlman, Daniel D. 2004. Knowledge Management Terms [Online]. Available: <http://home.earthlink.net/~ddstuhlman/defin1.htm>. [cited September 12, 2005]

REFERENCES

PERIODICALS

Chief Executive [Online]. Available: <http://www.chiefexecutive.net/Index.htm>. [cited September 12, 2005]

CIO [Online]. Available: <http://www.cio.com/>. [cited September12, 2005]

Computerworld [Online]. Available: <http://www.computerworld.com/>. [cited September 12, 2005]

Darwin [Online]. Available: <http://www.darwinmag.com/>. [cited September 12, 2005]

Harvard Business Review [Online]. Available: <http://harvardbusinessonline.hbsp.harvard.edu/b02/en/hbr/hbr_home.jhtml>. [cited September 12, 2005]

KMWorld [Online]. Available: <http://www.kmworld.com/>. [cited September 12, 2005]

QUOTATIONS

Bartlett, C.A., and S. Ghoshal. 1995. Changing the Role of the Top Management: Beyond Systems to People. *Harvard Business Review,* May-June 1995:132-142.

Davenport, T.H. Think Tank: The Future of Knowledge Management. *CIO,* December 15, 1995.

Drucker, P.F. 1995. *Managing in a Time of Great Change.* New York: Penguin.

Hedberg, B., and S. Jonsson. 1978. Designing Semi-Confusing Information Systems for Organizations in Changing Environments. *Accounting, Organizations, and Society,* 3(1):47-64.

Kransdorff, A. 1998. *Corporate Amnesia: Keeping Know-How in the Company.* Oxford: Butterworth-Heinemann.

Lincoln, A. 1862. Annual Message to Congress – Concluding Remarks. Washington, D.C., December 1, 1862.

Schlender, B. with B. Gates. 2003. The Vision Thing Again. *Business 2.0,* December 2003.

Zuboff, S. 1995. The Emperor's New Workplace. *Scientific American,* 273(3), 202-204.

BIBLIOGRAPHY

STRUCTURE AND RATIONALE

Because the published literature for Knowledge Management (KM) as it relates to the water industry by itself is not prolific, this bibliography incorporates KM characteristics, best practices, case studies, and metrics from other utilities and business sectors as well. By project participant consensus, the bibliography includes citations, abstracts, and keywords in order to increase its utility.

For ease of use, the 500-some references (which are comprised of articles, books, conference proceedings, and Web citations) are sorted alphabetically by author within major bibliographic divisions and subcategories, listed below. Placement of a reference within a given category is somewhat arbitrary; for example, a reference cited in "Aerospace Industry" could at times just as easily fit under "Lessons Learned." For the sake of brevity, each reference is cited only once.

A variety of engineering, legal, medical, and online databases were consulted in the generation of the bibliography, including ABI/Inform, Amazon.com, Business Source Premier, Compendex, LexisNexis, Library of Congress, INSPEC, and MEDLINE/PubMed.

Both the project team and research team recognized that the value of the bibliography as a resource would be increased substantially if the citations were presented in a searchable electronic file format. A CD-ROM containing the searchable electronic bibliography in PDF format is packaged in a slip inside the back cover of the report.

A word about Keywords: For the most part, the Keyword designations derived from the various databases are retained, partially to demonstrate the evolution in KM-related terminology over the 16 years represented by this bibliography, but also to aid the user in searching for additional citations from the literature.

KM Bibliography Divisions and Subcategories

The bibliography is sorted into five major divisions, with several subcategories contained within, as follows:

<u>Primers</u>

- Primers and Sourcebooks
- Best Practices
- Critical Success Factors
- Lessons Learned

<u>People, Process, Culture</u>

- Collaboration
- Communities of Practice
- Corporate Culture (including such terms as Change Management, Organizational Learning, Organizational Development)

Technology

- Content Management (including Customer Relations Management, Document Management, Drawing Management, Imaging, Records Management, and Web Content Management)
- Databases (including Business Intelligence, Data Mining, Data Warehousing, Knowledge Bases)
- Portals (including Intranet)
- Search Engines (including Enterprise Search)

Industry Applications and Studies

- Cross-Industry Case Studies
- Aerospace Industry
- Architecture, Engineering, and Construction Industries
- Automotive Industry
- Electric Power Utilities
- Legal Profession
- Medical Profession
- Military
- Mining and Petroleum Industries
- Utilities (Nonwater)
- Water Industry

Metrics, Models, Tools, and Techniques

- Metrics (including QualServe)
- Models, Tools, and Techniques

CONVERSION TABLE

SI EQUIVALENT UNITS

	To Convert		
Water/Wastewater Treatment Parameter	**From Customary Units**	**To SI Units**	**Multiply By**
Treatment Capacity	MGD	m^3/d	3.785 x 10^3
		or ML/d	3.785

ABBREVIATIONS

ACE	Annual Conference and Exposition
AMR	Automatic Meter Reading
AMSA	Association of Metropolitan Sewerage Agencies
AMWA	Association of Metropolitan Water Agencies
AWWA	American Water Works Association
AwwaRF	Awwa Research Foundation
BGD	billions of gallons per day
CAD	Computer-Aided Design
CIP	Capital Improvement Program
CoP	Communities of Practice
DCS	Data Control System
EPA	Environmental Protection Agency
ERP	Enterprise Resource Planning
GIS	Geographic Information System
HR	Human Resources
HTTP	Hyper Text Transfer Protocol
IMTech	Information Management and Technology
Inc.	Incorporated
IS	Information System
IT	Information Technology
K	thousand
KM	Knowledge Management
LDAP	Lightweight Directory Access Protocol
LIMS	Laboratory Information Management System
M	million
M.Eng.	Master of Engineering
MGD	millions of gallons per day
MS SQL	Microsoft Structured Query Language
NA	not applicable
N/A	not applicable
NACWA	National Association of Clean Water Agencies
NASA	National Aeronautics and Space Administration
O&M	Operations and Maintenance

OS	Operating System
PAC	Project Advisory Committee
PDA	Personal Digital Assistant
P.E.	Professional Engineer
P.Eng.	Professional Engineer
psi	pounds per square inch
RFP	Request For Proposal
SCADA	Supervisory Control and Data Acquisition
U.S.	United States
U.S.A.	United States of America
WEF	Water Environment Federation
WEFTEC	Water Environment Federation Technical Exhibition and Conference

IWA Publishing's authorised EU representative for General Product Safety Regulations is Diane D'Arras, 15 rue Duret, 75116 Paris, France, e-mail: safety@iwap.co.uk.

Printed and bound by CPI Group (UK) Ltd, Croydon, CR0 4YY
17/09/2026
02218793-0016